Hazardous Metropolis

Hazardous Metropolis

Flooding and Urban Ecology in Los Angeles

Jared Orsi

UNIVERSITY OF CALIFORNIA PRESS
Berkeley · Los Angeles · London

University of California Press
Berkeley and Los Angeles, California

University of California Press, Ltd.
London, England

Library of Congress Cataloging-in-Publication Data
Orsi, Jared, 1970–
Hazardous metropolis : flooding and urban ecology in Los Angeles / Jared Orsi.
p. cm.
Includes bibliographical references (p.) and index.
ISBN 0-520-23850-8 (cloth : alk. paper)
1. Flood control—California—Los Angeles. 2. Flood control—Government policy—California—Los Angeles. 3. Urban ecology—California—Los Angeles. I. Title.
TC424.C2 O77 2004
363.34′936′0979494—dc21 2002155797

Manufactured in the United States of America

13 12 11 10 09 08 07 06 05 04
10 9 8 7 6 5 4 3 2 1

The paper used in this publication meets the minimum requirements of ANSI/NISO Z39.48–1992 (R 1997) *(Permanence of Paper)*. ∞

I dedicate this book to my grandparents,
Raymonde and John, Elmer and Ogda,
who are the reasons I love Los Angeles.

Contents

Illustrations

PHOTOGRAPHS

Photographs follow page 74.

MAPS

Acknowledgments

In the course of researching and writing this book, I have been blessed with abundant help. Arthur McEvoy has been a personal and scholarly model for more than a decade, and it is to him that I owe the greatest thanks for anything that is creative in this project. Whenever recommending one of his favorite books or his foolproof model for writing, he often promised me, "This will change your life." My studies with him did. Another special thanks goes to Richard Orsi, a boundless source of editing assistance and research advice who has gone well beyond even fatherly duties to nurture this book and its author.

The book has also benefited from the contributions of many others. Mary Coomes, Bill Cronon, Colleen Dunlavy, Mark Fiege, Glen Gendzel, Lynne Heasley, Ari Kelman, Phoebe Kropp, Bill Philpott, Jenny Price, Sara Pritchard, Louise Pubols, Amanda Seligman, Rebecca Shereikis, Marlene Smith-Baranzini, Greg Summers, and Marsha Weisiger read drafts of this work and offered many useful suggestions along the way. Henry Binford, Turpie Jackson, Florencia Mallon, Mort Rothstein, Frank Safford, Michael Smith, and Robert Wiebe also contributed to this project indirectly by cultivating my intellectual development in general. One of the best things about working in the Colorado State history department is having Mark Fiege as a colleague; he read multiple versions of the manuscript, and our frequent conversations always spark new ideas. I have been deeply humbled over the last few years by how much I have come to look up to my younger brother, Peter Orsi, who copyedited the entire penultimate version of the text and whose writing advice greatly improved the manuscript. Finally, I owe a debt I cannot repay to a fabulous editor and historian and a dear friend, Graham Peck; it often seemed that he spent as much time with the manuscript as I did.

In the course of working on the project, I have had the privilege to meet many wonderful individuals who have taught me much about southern California environmental history. Dave Rogers took an early interest in my work, patiently explained the technical aspects of flood-control engineering, and opened to me his private collection of California civil engineering history and maps. Blake Gumprecht shared with me his vast knowledge of Los Angeles history and geography. Anthony Turhollow guided me through the materials at the Public Affairs Office of the Los Angeles District of the Army Corps of Engineers. Ron Lockmann was another invaluable contact at the corps, generous with his time and historical materials. Sarah Elkind was a constant intellectual comrade, always ready to talk L.A. history or to share research notes. And Jenny Price was a witty, insightful, and energetic partner on the many field trips we took exploring the banks and the bed of the Los Angeles River.

I also owe thank-yous to Geoff Barta, Peter Blodgett, Randy Brandt, Patrick Chan, Bruce Crouchett, Dennis Crowley, Bill Deverell, John Engemann, Irv Gellman, Bob Gottlieb, Wil Jacobs, Lewis MacAdams, Carol Marander, Chris McCune, Char Miller, Natalia Molina, Chi Mui, Sarah Pfatteicher, Peter Reich, Martin Ridge, Ray Sauvajot, Janet Sherman, Mark Stemen, Dace Taube, Linda Vida, Jim Williams, Paul Wormser, and Terry Young, who contributed in ways too numerous and varied to itemize. The book also bears the imprint of several scholars whom I have never met, but whose intellectual influence on me goes beyond what this book's frequent citations of their work can convey; among these are Stephen Jay Gould, John McPhee, and Charles Perrow.

Generous financial support was provided by the California Institute of Technology, Colorado State University, the Haynes Foundation and the Historical Society of Southern California, the Huntington Library, Northwestern University, the Society for the History of Technology, the University of California Humanities Research Institute, and the University of Wisconsin–Madison. The editorial and production staff at the University of California Press shepherded this project through the publication process, and three outside reviewers provided fresh suggestions that helped me tackle revisions.

I also want to acknowledge with gratitude the family and friends who provided meals and lodging, cars, camaraderie, Dodger tickets, and assorted other kindnesses that made my research trips to California thoroughly enjoyable: Suzanne Brown, Paul Cheng, Andor Czigeledi, Bonnie Hardwick, Beth Hessel-Robinson, Lauren Lassleben, Don and Suzie Orsi, Rae Orsi, and Barbara, John, and Mark Pesek. Finally, I am grateful for the unconditional and unwavering love from my family, Richard, Dolores, and Peter Orsi, and Jim, Nadine, and Matt Hunt. No words are adequate to convey my thanks to my mother, Dolores, whose love of reading I was fortunate enough to inherit and whose talent for writing I have always aspired to imitate. To my daughter, Renata, who was born five days after I submitted the revised manuscript to the Press, I am in debt for providing an irresistible incentive to meet publication deadlines. Finally, I can think of no better word than *companion*—in Latin, literally, one who shares bread—to describe Rebecca, and it is to her that I owe the greatest thanks of all.

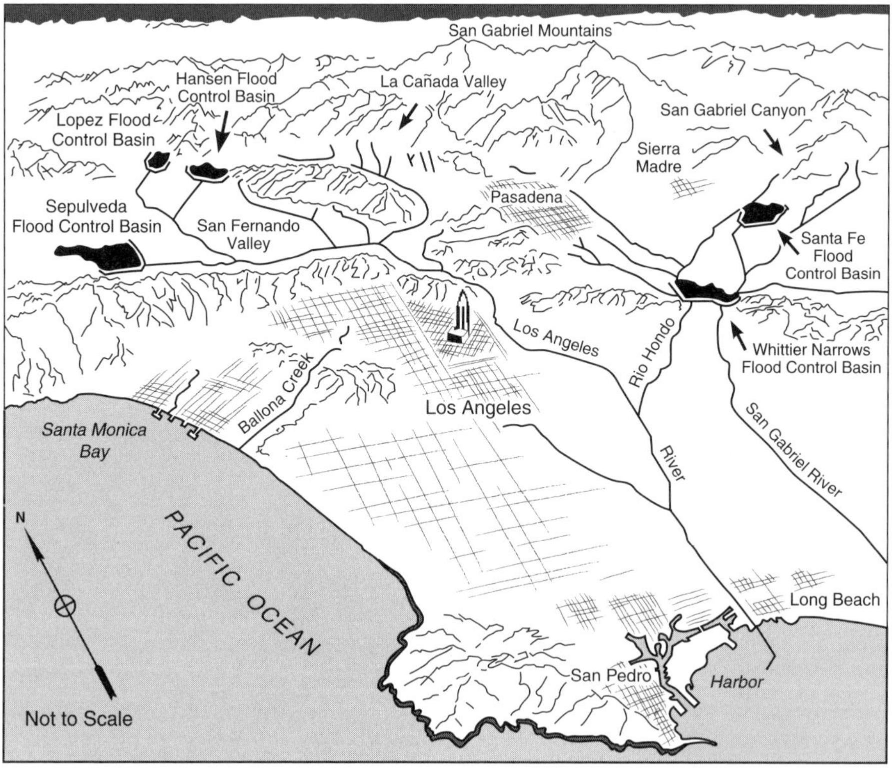

Map 1. Geographical features of the Los Angeles Basin. The metropolitan area that is the subject of this study is bounded on the east and west by the borders of Los Angeles County. Mountains as high as ten thousand feet in elevation crown the area on the north, and several lower ranges of hills partition the inland San Fernando and San Gabriel valleys from a broad coastal plain that slopes gently to the Pacific Ocean, the southern boundary of the study. The two principal river systems, the Los Angeles and the San Gabriel, lose more elevation in their fifty-mile courses to the sea than the Mississippi does in its entire route to the Gulf of Mexico. During the twentieth century, humans confined most of the rivers in concrete-lined channels. Debris basins were dug in the foothills to catch the boulders, sediments, and other debris the water carried out of the mountains. And five flood-control basins were built on the valley floors to trap flood water and funnel it slowly downstream during storms. (Map by Chris Phelps.)

PROLOGUE

Water in Los Angeles

A Portrait of an Urban Ecosystem

A winter storm rolled onshore at Los Angeles on 13 February 1980. A second storm followed a day later, then a third, and a fourth, and a fifth. A sixth storm brought the heaviest rains yet, swelling the Los Angeles River to its levee tops. Meanwhile, weather forecasters spotted a seventh storm brewing out on the Pacific. As water rose in the dark that night, the swamped electronic stream gauges were malfunctioning, and the technicians at the flood-control headquarters lost track of exactly how high the water was running. If the river were to spill over its walls, it would eat away at the levees from the landward side. They would crumble, and the torrents would gush into the adjacent neighborhoods. No one knew if the channels could handle one more storm. Fortunately, the rain stopped that night, and the seventh storm never materialized. When the sun rose the next morning, inspectors from the Los Angeles County Flood Control District found flood debris strewn atop the levees near the Wardlow Road overpass in Long Beach. Apparently, it had been a very close call.[1]

The near miss struck after six decades of flood control, during which engineers had redesigned the rivers of southern California.[2] Ever since the rivers inundated the city in 1914, conventional engineering wisdom had sought to replace the rivers' perceived disorderliness with the rationality of human artifice. Every time it flooded, the solution was to build bigger, stronger, and better structures—bulldoze some more channels, dam a river or two, armor the levees with another layer of protection—and each time, another destructive flood followed a few years later. In theory, at least, there were other ways to control the floods. Most of these would have involved regulating land use in one manner or another, and until the 1990s, none was considered more than half seriously. Instead, as one critic put it, the flood controllers reduced the problem of flowing water to an issue that "should only be dealt with by engineers and solely as an engineering problem."[3] Indeed, by 1980 it seemed the flood controllers had mastered the rivers. Water gathered behind small check dams in the mountains and sprawling flood-control basins in the valleys. It made turns to avoid harbors and other urban obstacles, and nearly every inch of its way to the sea it coursed over concrete beds between paved levees. This new hydraulic regime, the flood controllers calculated, would contain all but the greatest of floods.

The deluge of 1980, however, was not a great flood. In fact, the most extraordinary feature of the storm was its ordinariness. Flood-control experts estimated that a storm of that severity could be expected to occur on average about once every twenty-five to forty years. According to the engineers' calculations, such a storm should not have deposited any debris on the levee at Wardlow Road. Not only did debris top the levee, but a lake doubled in size; a flood-control channel crumpled while carrying less than its design capacity; and flowing mud smashed suburban foothill homes. The toll for this unextraordinary storm was $270 million in property damage and eighteen human lives.[4] With devastation so out of proportion to the size of the storm, the engineers recalculated. They raised the walls along the levees another few feet and pronounced Los Angeles safe again. The dialectic continues.

The storm, however, raises some heretofore unasked questions. Why, over the course of the century, did the engineering structures keep failing, and why did people keep building them? Put another way, why have bulldozers and concrete been so consistently appealing even though they have not always controlled the floods? Answering these questions leads simultaneously along two related paths of inquiry. Along one path unfolds the story of the flow of water in Los Angeles and the complex of forces that left water atop the levees at Wardlow Road and in other places it was not wanted. The second path invites us to generalize from Los Angeles to explore the ecological and historical structure of the urban places we inhabit today. The stories of both Los Angeles in particular and urban ecological structure in general begin in the water.

Water flows downhill. That much is simple. From there, however, things get more complicated. Laboratory studies show that even in a pipe, such factors as velocity, volume, gradient, obstacles to flow, and shape and roughness of the conduit lend stunning complexity to the flow of water. Consequently, many apparently simple questions have long confounded scientists. What makes gently flowing water suddenly burst into roiling billows, with eddies within calms within eddies? How can such turbulence be modeled? Why does something so simple produce such complexity? The flow of water outside laboratories is even more turbulent. So is most flood control. In the stream beds of the Los Angeles Basin, the shape of the conduits, the obstacles to flow, the roughness of the beds, and the volume, velocity, and gradient of water are determined by complex interactions of environmental processes, human activities, and, of course, the water itself. As a result, over the last century, as southern Californians have sought to impose order on their waters, they have discovered that their own hydraulic systems are just as disorderly as the rest of nature's.

THE PATH FROM SKY TO SEA: WATER'S JOURNEY THROUGH LOS ANGELES

The Sky

Much of the water that passes through Los Angeles begins its journey above the ten-thousand-foot peaks of the San Gabriel Mountains, north of the metropolitan plain and only fifty miles from the sea. Rainfall is unpredictable there. A rise of a few degrees in water temperature off the Peruvian coast can triple the annual rainfall in Los Angeles, and Alaskan air-pressure changes one day can send a tempest over the San Gabriels two months later.[5] This meteorological volatility renders the concept of a normal season meaningless in southern California. Annual rainfall fluctuates from four inches to forty, and it matches its historical average only about a fifth of the time.[6] Deluges often precede droughts, such as in the 1860s, when a thirty-day downpour was followed by three desiccated years that parched the land, starved the cattle, and ruined many a rancher. The water that does come falls unevenly. Meteorologists have long described storms in terms of expected return intervals. A ten-year storm, for example, is likely to occur once every decade, with a hundred-year storm coming, on average, once a century.[7] Statistically, the probability of a ten-year storm occurring in any given year is one in ten; for a hundred-year storm it is one in one hundred. Nothing could be more misleading, however—so misleading that some scientists recommend abandoning the concept altogether. Storms are not evenly distributed over long time frames, but come in bunches. Hundred-year storms may occur several times in a decade or even in a single year. Moreover, in Los Angeles, as in much of the rest of the western United States, climatic records are short, and the scientific guesswork that goes into the estimates is considerable; as a result, the prediction of storm frequency is often wrong. Storms of a size once thought to be exceedingly rare often turn out to be much more frequent events.[8] Even a moderate gale produces bands of weather, called cells, that drop large amounts of water on particular locales. A given squall might be classified as a ten-year event over the entire region, but over one mountain or canyon, it might deliver hundred- or even thousand-year rains. Thus the basin can receive hundred-year storms at different locales several times in a decade.[9]

On New Year's Eve 1933–1934, wet tropical air brought light rains to the southern California coast. Moving inland, the warm air met an eastern cold front. The cold air dove, lifting the warm air mass before it reached the mountains, and an intense cell developed over the foothills. It just so happened that the clouds burst over hillsides that had already been saturated by another storm two weeks earlier and had also been burned by fire three weeks before that. Water rushed down the denuded slopes, gathering the fire's rubble into globs and spreading them over the recently developed foothill suburbs.[10] Forty people died, crushed or drowned by flowing mud. Fire, rain, a warm front, a cold front, and then more rain—each had happened before in southern California, but never, since the settlement of the foothills, had they all happened together. City and climate interacted to produce an urban ecological disaster. It would not be the last time.

The Mountains

Most often, however, the clouds burst over the higher elevations. Inland-moving air rises as it meets the hills, depositing some moisture on the foothills and saving most of it for the summits. Some of the water evaporates. Some soaks into the ground. Some runs over the surface into gullies that feed into canyons. Tectonic movement has lifted the mountains for hundreds of thousands of years, but water has worn them down at the same time. Along the fractures in the rock, where the forces have lifted the mountains unevenly, the surface material is easily worn off, and water, seeking a downward path of least resistance, invariably follows these channels.[11] Water flows down. Mountains push up. Together, the two unrelated forces convey the drainage waters of the San Gabriel Mountains along the weakest, most fractured paths possible.

Consequently, water often flows precisely where the geology will not support flood-control engineering. In 1924, the Los Angeles County Flood Control District planned to build the world's tallest dam in San Gabriel Canyon to store the floods as insurance against the dry spells. The scheme proposed to expand the cost, scale, and purpose of flood control, even though engineers knew little about the canyon's geology, and voters had only barely supported comparatively modest previous flood-control measures. The impossible project got underway only because the 1924 dam bond election coincided with both a real-estate boom that made the project economically feasible and a drought that made it politically popular. After a year of excavation, however, the sides of the canyon at the construction site collapsed. There was no bedrock on which to build the great dam. It also turned out that the contractors had known this and had defrauded the public of hundreds of thousands of dollars. Consequently, the discredited district could not marshal votes for flood-control bonds even in the aftermath of the catastrophe on New Year's Eve 1933–1934. The geological knowledge, the Flood Control District management, and the rock itself were all faulty, and together, they determined whether or not the world's tallest dam would block the water in San Gabriel Canyon.

The Foothills

Water flowing through the canyons gathers debris. The heaviest of this rubble remains high up in the canyons, where masses of soil, rocks, plant matter, and fire debris pile up, while seasonal streams carry the lighter silt downhill out of the canyons and deposit it on the foothills. The foothills themselves are products of this process, as hundreds of thousands of years of deposition have built up gently sloping hills of debris, called cones, that fan out from the mountain front. In dry years, the streams leaving the canyons course over the tops of these cones in shallow beds, which they fill with silt, raising them nearly to the level of the adjacent ground. In wet years, however, when storm cells pass over burnt slopes and silted beds, rainfall and runoff soak the masses of rubble that lie back up in the canyons. Once saturated, the masses inch forward and then flow faster and faster, eventually picking up tree trunks and boulders and crashing down the canyons. When they reach the cones, the masses are at their most unpredictable. They easily overflow the shallow, silted stream beds, and when they do, nobody nearby is safe, because it is impossible to guess where the debris will go.[12]

Engineers have only partly learned how to control this debris. Beginning in the 1930s they dug stadium-sized pits, called debris basins, to capture the muck at the mouths of the canyons and allow clear water to drain out the bottom. In 1978, as in 1933–1934, another weather cell poured its rain onto recently burned foothills. As sludge gushed into the debris basins, boulders and mud clogged the drains. To the amazement of engineers, the clogged drains and fluid dynamics of liquefied mud turned some of the debris basins into giant "flip buckets," as one Flood Control District official put it, that propelled waves of mud over the spillways and onto homes even before the basins had filled to capacity.[13] Thus the 1933–1934 New Year's Eve tragedy recurred, only this time, infrastructure failures were accomplices to the atmospheric disturbances and suburban growth. Despite the best technical efforts, water still flowed destructively.

The Plain

Monster debris flows, however, are rare. Usually, the water stays in its beds until it reaches the coastal plain, though even there a riverbed is not always a stable thing. Two major river systems drain the Los Angeles Basin, the Los Angeles River on the west and the San Gabriel River on the east. They are joined by the a third waterway, the Rio Hondo, a distributary of the San Gabriel that delivers its waters to the Los Angeles River. Before the twentieth-century flood-control efforts, these rivers and their tributaries would flow in their beds until sediment buildup barricaded their paths and rains filled the channels to capacity, forcing the water to jump its banks and strike out in a new direction—and not just by a few yards here or there. In the late nineteenth century, recently arrived immigrants from the eastern states did not understand this volatility of western streams, nor did they pay much attention to the Mexican residents who did. Those old-timers recounted an 1825 flood that had covered the entire countryside with water and changed the course of the Los Angeles River. Before that event, the river had bent westward along what is today Washington Boulevard and flowed out to Santa Monica Bay. In the deluge of 1825, the river carved a southward path to San Pedro Bay, some twenty miles from the old outlet.[14] Dismissing these accounts, the newcomers built right up to the edges of—and sometimes right in—the usually dry riverbeds. After the Los Angeles River again went westward in 1884, flowing into Santa Monica Bay, the Los Angeles City Council tried to legislate permanent boundaries for the channel and authorized a rail company to build a levee along one bank. A few years later, however, the river escaped its legal confines and took a new path over adjacent farm fields, prompting a lawsuit to determine who would be liable for the water's delinquency.[15]

Today, concrete, steel, and round-the-clock monitoring keep the rivers in their beds. Whittier Narrows Flood Control Basin stands amid urban sprawl, in the heart of what used to be a tangle of silt- and vegetation-choked channels, where the San Gabriel River used to choose a path—or paths—to follow. This long, low barricade, one of five in the county, provides the river two outlets. The majority of the basin is usually dry, but when waters rise during storms, the Flood Control District's command center buzzes with activity. Weather reports and river data pour in, telling operators how much water is coming, when, and from where. With this information, the operators decide how much water to keep behind the dam and how much to release downstream. But the pavement that covers the region causes water to concentrate quickly, and the unpredictability of Pacific storms sometimes leaves flood fighters guessing. Dam operators have little time to consider this imperfect data as they make the weighty decisions that could determine whether or not the levees hold.

So far the hyperregulated flood-control basins have not failed their tests. But in past deluges, the rivers have formed enormous lakes south of the Narrows, and it is not inconceivable that the Los Angeles River might one day try to reclaim Washington Boulevard. Since the close call near Wardlow Road in 1980, the U.S. Army Corps of Engineers has feared that rivers overtopping the levees might send water gushing over eighty-two square miles that are home to five hundred thousand people, many of whom have moved onto the flood plain believing that the concrete levees protect them. Such an event would likely do more than two billion dollars of damage.[16]

The Harbor

Assuming it stays between its levees, at the end of its journey the water spills into the ocean east of the Los Angeles and Long Beach harbors. Although today the two ports form a single harbor, which is the nation's busiest, their history is something of a metaphor for southern California's reputation as a place where grand appearances sometimes veil a lack of substance. The retired British ocean liner, the *Queen Mary,* now a restaurant and hotel, lies anchored next to Long Beach Harbor, pointed upstream as if preparing to sail right up the Los Angeles River. "It seemed a nice Southern California touch," a journalist once quipped about the scene, "a ship without engines steaming up a river without water."[17] Like the *Queen Mary,* the harbor itself is hardly what its appearance suggests. It is no natural harbor. In dry years, the area that is today the harbor was the occasional outlet for what little water the sluggish Los Angeles and San Gabriel rivers delivered to the ocean. In wet ones, it was the repository for loads of mountain silt, which the tides would scour out to sea. There was nothing inherent in this location that required southern Californians to put their harbor on this site, at the mouth of the county's two main drainage systems. In 1897, this location won out over others for political reasons, following nearly twenty years of bickering among southern California business leaders. As the Army Corps dredged the mudflats to create a deep seaport in the first decade of the twentieth century, no large floods struck to warn of the threat to the port. Then, after a 1914 torrent, sediment choked the harbor for days, spurring southern Californians to launch the countywide flood-control efforts that continued throughout the twentieth century.

From sky to sea in Los Angeles, water passes through an urban ecosystem. In that system, the political, social, economic, cultural, and physical features of the city have joined climate, geology, biology, and topography to determine when, where, and how water flows. That makes it urban. Meanwhile, as in even the simplest ecosystems, there are far more factors influencing the flow of water than engineers can possibly take into account, and these factors interact in ways that engineers cannot possibly predict or quantify. Small causes explode into big effects. Unrelated factors interact. Coincidental timing leads to unlikely events. Processes in the system generate their own demise. All this makes it ecological.[18] From the vantage point of those who would quantify, predict, and redesign urban ecology, it is a very disorderly system.

But it is not a unique system. Human-set wildfires engulf New Mexico towns. Central American mudslides swallow hillside shanty communities. Manhole covers explode from the streets of Georgetown. Mexico City earthquakes unmask years of corruption that allowed shoddy apartment construction. Tornadoes uncannily seek out American mobile home parks. Pavement in Chicago suburbs prevents rainwater from getting into the ground, threatening those cities (whose annual precipitation matches Seattle's) with water shortages of a severity usually associated with the arid West. Everywhere one looks, it seems, the boundaries between the wild and the urban blur, often with tragic consequences for those who occupy cities believing they have escaped nature's disorder. In recognition of these complicated boundaries and the complex of problems they pose, the U.S. National Science Foundation in 1997 invited urban proposals for its Long Term Ecological Research Program, which had been funding comparable long-term studies of wildlands since 1979. The NSF, with its growing interest in urban ecology, is implicitly responding to half a century of massive urbanization worldwide. Increasingly we humans live in these sorts of landscapes. We need to understand them.

In particular, we need to understand that urban ecosystems are not merely transitory systems. Historically, conventional wisdom in Los Angeles and elsewhere has treated these sorts of landscapes as oddities—worlds in transition from the messiness of nature to the clean predictability and regularity of human artifice. If chaos prevailed, people assumed, it was only because humans had not yet ironed out all the wrinkles. Through this lens, every flood appeared to be an indication of the need for more bulldozers and more concrete, not a warning that other types of solutions should be added to the defenses. With hindsight, however, Los Angeles history belies the transitory nature of urban ecosystems. For 120 years, elements of the city's physical and human landscapes have intertwined and united to direct the flow of water along unpredictable and occasionally destructive paths. Clearly this is not a way station on the path between disorder and order. What explains the apparently accidental combinations that cause these factors to work together to make water flow in such destructive ways? Is there some structure that links the storm cells, rock fractures, clogged drains, unstable river channels, inadequate levees, urban politics, and ersatz harbors—some explanation, that is, other than chance?

This book attempts to work out that structure. It explains the urban ecosystem historically—what brought it into being, what makes it work, and how it changes. The structure of urban ecosystems, it turns out, is exceedingly complex. It is this complexity that makes bulldozers and concrete so irresistible while rendering other strategies virtually invisible. And it is because of this complexity that urban artifice mirrors and incorporates all the disorder of the rest of nature. Successful flood control requires not only measuring rainfall, debris bulk, and concrete strength, but also considering politics, economics, social relations, and cultural values—and understanding how these unquantifiable factors interact with the technical aspects of the problem.[19] In short, flood control in Los Angeles is much more than solely an engineering problem.

CHAPTER 1

City of a Thousand Rivers

The Emergence of an Urban Ecosystem, 1884–1914

A TERRIBLE AND GRAND OLD RIVER

In the winter of 1884, the *Los Angeles Express* complained that the usually trickling Los Angeles River had turned into a "terrible and grand old river."[1] By early February, three months of rain had so moistened the ground that it could absorb little more. The *Express* wondered on 7 February if the rain would ever stop, but the downpours continued almost without pause for another month. Water began to gather, forming ponds and then lakes, and then it spread across the countryside. South of the town of Los Angeles, where the coastal plain stretches twenty miles to the Pacific Ocean, the river ran several miles wide, and people could row their boats seven miles between the communities of Compton and Artesia. After catfish and carp were netted outside a blacksmith shop in the town of Norwalk, one wag put out a sign saying No Fishing. Houses and crops washed away. People and livestock drowned, and the cascading water smashed buildings, bridges, barns, and fences. Many old-timers recalled it as the greatest flood in memory.[2]

The terrible and grand old river, however, provoked comparatively little response from people at the time. With the exception of some damage done to the few cultivated or inhabited lands, the currents spread harmlessly for miles over the plain, deposited their silt loads, and soaked into the ground or flowed out to the sea, leaving little destruction in their wake. Aside from attempts by the town of Los Angeles to build protective levees and scattered efforts by property owners to wall off their lands from the streams, people took few steps to prevent future recurrences of the event. The precautions they did take were for the most part small-scale, localized, and private. Some people even counted the floods a blessing. As Norwalk's James Hay noted, the fertile alluvial sediment turned "black alkali" into "some mighty good ground." F. A. Coffman of Rivera remembered, "We really believed it did good instead of damage."[3] In a chronically thirsty land, people welcomed water—however it came. Frequently, flood was how they got it. Inundations of similarly epic proportions had visited southern California throughout the century, most recently in 1862, 1867–1868, and 1881, and they would do so again in 1886, 1889, 1890, and 1891. The terrible and grand old river of 1884 was merely one in a series of nineteenth-century deluges that were dramatic, inconvenient, and occasionally destructive but caused little change to where and how people lived.

The same cannot be said of the flood of 1914. In that deluge, crops and homes washed away; ships mired in the silt that the swollen rivers delivered to the harbor; and virtually every bridge in the county was ruined. The region was marooned for days, cut off from communication with places outside the area. As water spread over the land, one newspaper proclaimed Los Angeles "the city of a thousand rivers."[4] Despite their similarities, the two deluges could not have differed more in their effects. The 1914 flood did considerably more property damage and caused a greater public stir, as citizens immediately called for comprehensive flood control and set up an agency to meet such demands.[5] Remarkably, however, the 1914 flood that did so much damage and elicited such a swift public response was estimated to be the smaller of the two floods by 30 percent.[6]

The smaller flood provoked such a determined response because it struck a radically different ecosystem. Of course, the landscape in 1884 was by no means pristine. Livestock had grazed the native grasses nearly to extinction, and invasive European plant species had taken root; farmers had plowed fields and dug irrigation ditches; and dirt roads and steel rails connected the region's scattered settlements. But much of the old hydrology remained intact. Streams still carried sediment out of the mountains and still overflowed their beds during storms. Thickets slowed and deflected the torrents, spreading them over the flatlands to sink in or work their way to the sea, depositing their silt as they proceeded. Ocean tides carried away the silt that made its way to the mouths of the rivers. Between 1884 and 1914, however, explosive urbanization turned the region's collection of towns, farms, and open space with a widely dispersed population of thirty-three thousand inhabitants into a metropolis of half a million. As they built a city, southern Californians also built a new ecosystem, in which environmental perceptions, commercial activities, ethnic relations, and physical elements of the urban landscape had as much influence on the flow of water as climate, topography, and geology did. These urban features enabled smaller floods to do more damage. Thus, between the floods of 1884 and 1914, southern Californians constructed the city of a thousand rivers.

WHERE NATURE HELPS INDUSTRY MOST

The letterhead of the Los Angeles Chamber of Commerce in 1915 read "Los Angeles—Where Nature Helps Industry Most." That phrase reflected a four-decade-long process through which southern Californians developed a belief that the region's environment, its climate in particular, was especially nurturing to human health and industry. This enchanted land, it was said, would cure illness, lure tourists and immigrants, grow the most amazing crops, and provide a bountiful resource base for the burgeoning metropolis. Proliferating after the drought of the 1860s, such rosy portraits launched southern Californians' romance with their climate, a romance that grew over the next several decades and elicited outrageous exaggerations and misrepresentations. Rumors spread of two-hundred-pound pumpkins, seven-foot cucumbers, and nineteen-foot tomato vines growing in the warm weather and fertile soils. The air, physicians alleged, could cure any ailment: pneumonia, tuberculosis, malaria, old age, insomnia, scarlet fever—the list was endless. Southern Californians invented a climate.[7]

This overestimation of environmental beneficence helped to make the 1914 storm so disastrous. The idyllic images attracted hundreds of thousands of immigrants, who for the most part were both ignorant of the region's violent flood history and condescending toward the Mexican inhabitants who did know about it. The immigrants were also city builders who aimed to turn nature's bounty into metropolitan prosperity. As a result, they came to believe the myths themselves. If they did not believe literally in every two-hundred-pound pumpkin tale, they at least accepted the general substance of the myth of environmental beneficence. The urban institutions they created propagated this belief. Municipal law even codified it as the city of Los Angeles declared an official channel for the Los Angeles River. The metropolis the immigrants built manifested both their confidence in the environment's benevolence and their ignorance of its potential violence.

Newcomers to southern California had not always shared the chamber of commerce's optimism, however. Since the first Spanish explorers mixed glowing reports with descriptions of starving sailors, impoverished natives, and barren landscapes, southern California's record as a promised land was inconsistent at best. The Franciscan priest Juan Crespi, who accompanied the 1769 expedition that established California's first permanent European settlements, said the site that later became Los Angeles had "good land" and "all the requisites for a large settlement." He also noted, however, the evidence of drought and "great floods" and "earth tremor after earth tremor, which astonishes us." Mission San Gabriel, founded three years later, had to be moved several times as the Franciscans sought a spot close to the water supply of the shifting rivers but out of the reach of their floods. Beset by droughts, floods, and crop freezes, the missions and pueblos sometimes struggled to get by. The often romanticized hide-and-tallow trade that followed the mission period supported but a few *rancheros* with lavish lifestyles and a larger, poorer population of Indian and Mexican workers. Boston writer Richard Henry Dana, who visited Los Angeles's port at San Pedro in the 1830s, called it "the hell of California." As late as the 1860s, southern California still had not yet solidified its reputation as a Garden of Eden. "Nature," one traveler wrote, "is obstinate here and must be broken with steam and with steel. Until strong men take hold of the State in this way and break it in . . . its agriculture will be the merest clod whacking." Sometime in the early 1860s, Ross Browne traversed the state on horseback as an agent of the U.S. government. He reported that Los Angeles County was a barren desert with an uncertain future. It would not be developed for many years, he predicted.[8] Many agreed with Browne. Certainly no one during the great drought of the 1860s could have convinced a *ranchero* whose cattle were dying of thirst and whose lands were being auctioned to pay debts and taxes that southern California's climate was benign.

By the late 1860s, however, more optimistic views emerged. When Browne returned a few years after his first visit, his tune had already changed. Upon seeing corn tall enough to block the sun from his stagecoach window, he wrote back to Washington saying that he wished to contradict his earlier report. Another traveler marveled at the "beauty perceived through the droughts" and the "embroidered and flowered geometry of gardens."[9]

These glittering reports, in combination with the arrival of transcontinental railroads in 1876, 1881, and 1886, attracted easterners in droves. In 1873, a group of self-styled "warm blooded and adventurous persons who could not endure the frigid cold of Northern winters" formed the California Colony of Indiana, a band of midwesterners who settled in southern California. One of the settlers, Thomas Balch Elliott, described the region as the "gem of the world as to climate, health, production, softness and exhilarating air." It was "picturesque and charming," with "soil of unexampled fertility" and the "rarest climate in the world." He even praised the "pure soft cold and abundant" waters of the region.[10] As such portraits filtered east and the railroads came west, drought turned into perpetual sunshine.

Best of all, it seemed, southern California was free from environmental hazards. "Los Angeles," a newspaper editorial observed in 1914, "has experienced nothing . . . that may be termed a disaster. . . . We have been practically immune from the spite of nature." One visitor wrote, "The visible wrath of God is not to be found here: no one ever froze or roasted to death." In sunny southern California, where crops grew large, invalids got well, and settlers grew rich, disasters were unthinkable.[11]

This belief in the absence of natural disasters made sense at the time. Climatic flukes, demographic peculiarities, and ethnic prejudices combined to obscure all evidence to the contrary. The years 1891 and 1914 bracketed one of the longest floodless periods in recorded Los Angeles history. With the exception of some occasional flooding on the San Gabriel River that did a small amount of localized damage, no major deluges struck during these years. The period also witnessed astounding immigration that brought from the eastern United States hundreds of thousands of people who had heard glowing reports about the salubrious climate from newspaper articles, railroad advertisements, real-estate handbills, and tourist accounts. The trainloads of immigrants who came during these years had no experience of the great floods earlier in the nineteenth century.

The Germain family, for example, settled in the San Gabriel Valley, east of the city of Los Angeles, in the 1890s. There they cultivated 130 acres of feed for their dairy cattle and 30 more acres of walnut orchards. When they first arrived, the San Gabriel River flowed in a deep, narrow channel about a mile from their home. They never thought that it would flood their lands. In a localized 1911 flood, however, overflowing water washed away their alfalfa fields, farm implements, two barns, and improvements they had made to their dairy. The damage cost them twenty-five thousand dollars. The flood of 1914 took nearly all they had left.[12] The Germains' experience of buying high-and-dry land only to have it flooded in later years was not uncommon. At a time when three-quarters of all Los Angeles–area residents had not been born there, few of them understood the potential for flooding.

Those who did were often at the margins of society. Many of them were older, and as John L. Slaughter, who had lived in Los Angeles for fifty-three years, complained in 1914, "The people that are coming in here in these later days think that the old-timers do not know anything of the floods." [13] The primary hindrance to the spread of flood knowledge, however, was ethnic prejudice. In 1825, when the Los Angeles River changed directions, California had been part of Mexico. Decades later, stories of the great inundations were still well-known among the Mexican community of Los Angeles.[14] Parents and grandparents passed their memories down to younger generations, so that as late as the turn of the century, even those who had not lived through the deluges knew of these events.[15] As the Mexican community built up a store of flood memories, however, its social standing was eroding. With the American migration from eastern states, Spanish-speakers constituted a dwindling percentage of the southern California population and occupied ever-lower levels of the labor force.[16] And while people of Mexican ancestry held political office at the state and local levels in the decade after the Gold Rush, their numbers steadily declined thereafter.

The easterners dominated not only the region's economy and politics but its culture as well. They founded the newspapers and civic organizations and other institutions that invented and promoted a romantic history for southern California and coupled it to the benign climate myth.[17] For the most part, the newcomers disdained the Spanish-speaking locals but at the same time romanticized California's past as a colorful though primitive era of lavish Spanish *ranchos* and peaceful Franciscan missions.[18] Neither view inspired the city builders to put much stock in the flood tales. Some found the stories that the Los Angeles River had once flowed westward impossible to believe. Others dismissed the stories as the talk of "old Mexicans." Rarely did such stories change the immigrants' behavior. The Germains, for example, heard about floods from a Mexican man whose mother had seen water covering the San Gabriel Valley. On one occasion, she had told him, she spent the night in a tree to escape raging water. But the Germains nevertheless expressed surprise when they were flooded out in 1911 and 1914.[19] Until they experienced flooding themselves, it was easy for people like the

Germains to associate such stories with a quaint and colorful Spanish and Mexican past that had little relevance to their own lives in the dry, booming present. Thus the arrangement of both knowledge and power in Los Angeles created a society in which the people with the most influence over the production of culture had the least understanding of the region's environmental past, while those with the best memory had the least power. A climatic fluke, rapid immigration, and ethnic prejudice fragmented the society's collective memory.

If they lacked memory, however, the immigrants had a clear vision of the future. Flooding lay far from the minds of these city builders, who were intent on bringing water from afar, constructing a harbor, and developing industry and real estate. The thousands of people who, like Thomas Balch Elliott and his companions, flocked to southern California in the late nineteenth century seeking health and prosperity created the urban institutions that transformed southern California into a metropolis.

The city-building careers of those who came to southern California for its pleasant climate are impressive. Elliott and his company founded the city of Pasadena. Groups of health seekers founded the nearby towns of Monrovia and Sierra Madre and played important roles in developing the areas around San Diego, Santa Barbara, Ontario, San Bernardino, and Banning and desert regions like the Salton Sea area, as well as numerous resorts. Charles Dwight Willard, a tubercular midwesterner who came to southern California in the early 1880s, resuscitated the defunct Los Angeles Chamber of Commerce, helped lead the movement to secure a harbor for Los Angeles, established the local branch of the Municipal League (an influential Progressive Era reform organization), and founded the booster magazine *Land of Sunshine,* which he later sold to Charles Fletcher Lummis, another health seeker, who had walked to Los Angeles from Cincinnati in 1884. One of the most tireless promoters of the city and its romantic past, Lummis edited the city page for the *Los Angeles Times* and founded historic preservation organizations and the Southwest Museum. Also arriving in the early 1880s were Harrison Gray Otis and his future son-in-law Harry Chandler (who came to Los Angeles to cure his weak lungs). In addition to publishing the *Times,* the two developed real estate, dominated Republican Party politics, advocated harbor construction, and thwarted the area's budding organized labor movement. Rounding out the list of health-seeking city builders was Abbot Kinney, who converted a marshy strip of coast into a tourist community of beaches, canals, theaters, piers, concert halls, and hotels, which he called Venice of America. He also served as president of the Forest and Water Association of Los Angeles County.[20] By attracting people like Elliott, Willard, Lummis, Chandler, Otis, Kinney, and thousands of others, the perception of the region as an environmental utopia was instrumental in building the city, for it was this rush of people that brought to southern California the capital that established the agriculture, industry, municipalities, infrastructure, civic organizations, cultural institutions, newspapers, recreational opportunities, and scientific societies that distinguished turn-of-the-century Los Angeles from its pastoral past. These institutions would lay the groundwork for the obsession with development that would drive the region's economy for the next century.

Another of these health-seeking, city-building immigrants was the historian James Miller Guinn. In 1883, he helped found the Historical Society of Southern California, one of the many civic organizations that emerged during the urban growth of the 1880s. In an 1890 article, which he entitled "Exceptional Years," he recorded a history of abundant natural disasters, yet offered that history as proof of southern California's magnificent climate. He found evidence of bad flooding in 1811, 1815, 1822, 1825, 1832–1833, 1842, 1851–1852, 1859, 1862, 1867–1868, 1873–1874, and 1883–1884. Although he recounted that in the earliest floods rivers had abandoned their beds, lakes had dried up, forests had died, crops had washed away, and sand had smothered fields, he portrayed recent floods as benign. Floods, he concluded,

> like everything else in our State, can not be measured by the standard of other countries. We are exceptional even in the matter of floods. While floods in other lands are wholly evil in their effects, ours, although causing temporary damage, are greatly beneficial to the country. They fill up the springs and mountain lakes and reservoirs that feed our creeks and rivers, and supply water for irrigation during the long dry season. A flood year is always followed by a fruitful year.[21]

Thus Guinn turned the record of natural disasters upside down. For Guinn, the legendary cataclysms affirmed the romance and color of California's past, even as he maintained that such "exceptional" events were benign in the present. Guinn, like many southern Californians, saw the region's flood history as a record of colorful aberrations.

By the 1890s, southern Californians not only believed in their environmental fantasy; they had also codified it in municipal law. After waters overflowed in Los Angeles in 1884 and 1886, the city arranged for a railroad company to construct a flood-protection levee through the center of town. In 1888, the city approved the completed structure, a portion of which ran through the middle of the mostly dry stream bed, and declared it the "official" western bank of the river in accordance with the 1886 city ordinance that defined the banks of the river and granted the railroad company the land on which to build its tracks and levees. During a storm in January 1890, water thundered down the channel, swelling the trickling river. The levee deflected storm waters over the river's eastern banks and onto farmland for several miles, washing away soil, depositing sand and boulders, and destroying more than nine hundred acres of cropland. The farmers sued the railroad. Despite abundant testimony showing that the river had historically shifted channels and frequently overflowed, the California Supreme Court, which heard the case on appeal, considered flooding to be "occasional," "extraordinary," and "unusual," and refused to hold the company liable. A structure in the middle of the bed of such a river, the justices ruled, was not "inherently, and according to its plan and location, a dangerous obstruction to the river such as ordinary prudence would have guarded against."[22] Thus the exceptional-years thesis found its way into even the region's legal framework.

Guinn's exceptional-years thesis and the state supreme court decision exemplify southern Californians' response to floods around the turn of the century. As long as the deluges were generally perceived to be exceptional and local, no comprehensive program to combat them emerged. Flooding between 1889 and 1891 did spur Los Angeles County to commission a team of engineers to report on flood control. That project, however, died on the vine, as drought set in for the rest of the decade, and the city builders turned their attentions to getting more water, not disposing of what they already had. After localized overflows along the San Gabriel River in 1911, citizens petitioned the County Board of Supervisors for flood protection. The board commissioned another engineering report, but again little came of it.[23]

The limited flood-control projects that did get underway were small-scale and either private or quasi-public. In the 1880s, individual towns, private landowners, and railroad companies diverted waterways and built barricades between their properties and the riverbeds. The most elaborate efforts were the seven small flood-protection districts that were formed around the turn of the century. State law allowed landowners to form such protection districts in order to tax themselves and cede their own land to the district to build levees. As in the case of the Santa Fe Railroad Company's levee in the Los Angeles River bed, however, one area's flood control often produced another's flood. Levees along a given stretch of channel so aggravated the flood risk downstream that landowners on occasion had to hire armed guards to patrol their works and protect them from sabotage by angry neighbors.[24]

Further limiting the effectiveness of localized flood-control efforts was the tendency of rivers to change courses. In 1884, the Santa Fe Railroad led a group of landowners who diverted the San Gabriel River into an old channel it had previously followed. It stayed there until 1891, when another flood forced it back out. Such events not only damaged the flood-control works but also occasionally left them protecting channels in which water no longer ran. Some landowners refused to cooperate with localized efforts to control rivers, because they feared investing capital on improvements only to find their project destroyed or standing high and dry after the next storm.[25] Effective flood control was impossible under such conditions.

Thus, between the 1870s and the 1910s, southern Californians invented a climate and built a city based on that misrepresentation. Those efforts, however, eventually proved unsustainable. Developments such as railroad levees in riverbeds created new ways that water could flow destructively at the same time as they stimulated the urban growth that increasingly depended on water staying in its channels. To make matters worse, climatic cycles and the arrangement of society in Los Angeles thwarted flood-control efforts. Power was divided among competing and often antagonistic parties; flooding was usually localized; river channels would not stay in one place; and a twenty-year floodless stretch evaporated what little political will overcame these factors. Thus political structure, regional hydrology, and public imagination of nature and history conspired to leave Los Angeles with few defenses against the flood menace. In the 1880s, Los Angeles was small enough to escape the catastrophic consequences of building a city on a floodplain. During the next two decades, however, southern Californians invested much in constructing a hazardous metropolis.

IN THE PATH OF THE FLOODS

Despite the claims of the chamber of commerce and others about the abundant blessings that the physical environment had bestowed on Los Angeles, nature did not do this without humanity's helping hand. In transforming southern California from a pastoral setting into a metropolis between 1870 and 1920, people remade the land. They fashioned a

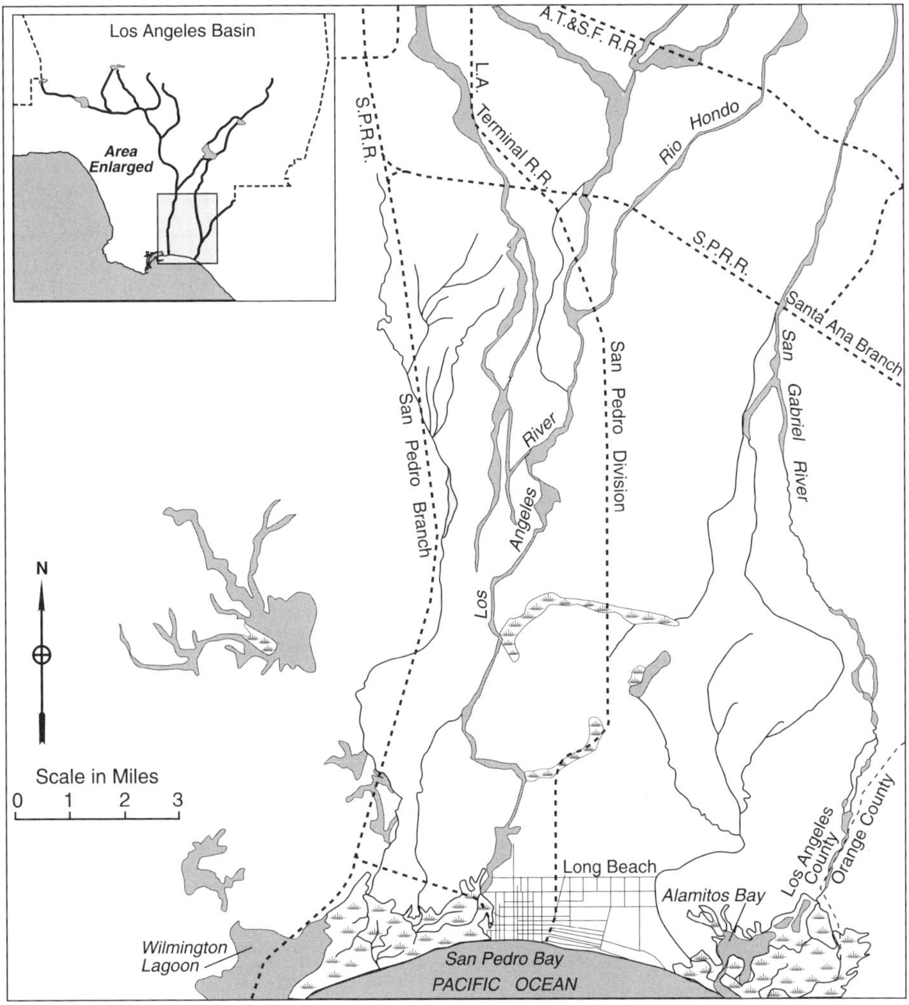

Map 2. Los Angeles coastal plain, circa 1894. Historically, southern California's streams flowed over the coastal plain in braided, frequently shifting channels before sinking into the wetlands around Wilmington Lagoon, San Pedro Bay, and Alamitos Bay, the future sites of southern California harbors. (Map by Carol Marander.)

harbor out of mudflats and turned grasslands and thickets into farms. Together, the rise of commercial agriculture and the construction of the harbor represented a substantial investment that took place in the paths of the old floods. As city builders rearranged the environment they had encountered in the 1870s, they inscribed their belief in a floodless climate right onto the landscape in the form of croplands growing in overflow zones and an eleven-million-dollar harbor at the mouth of silt-laden rivers.

The County's Greatest Single Asset

In April 1899, twenty thousand people gathered at Los Angeles Harbor to cheer the pouring of the first trainload of stone for a new breakwater.[26] *Harbor* was actually a rather generous term for the site. When the first ship sailed into it, the port was little more than a collection of tidal mudflats and sloughs. Even in the 1890s, ships still had to unload at sea and carry their cargo ashore on skiffs. The breakwater, however, heralded a busy commercial future. The stone poured that spring represented an early step in a decade-long dredging and filling project that would make the former marsh the busiest port in the nation.

The harbor was the most visible example of the urban boom that took place between the 1870s and the 1910s. As channels, jetties, breakwaters, and wharves were assembled into a port during the first decade and a half of the new century, business in the harbor district boomed. The tonnage of goods passing through the port surged from 200,000 in 1900 to 1.7 million in 1910, while the total value of the goods rose from two million dollars to more than a hundred million during the same period. In the lumber trade, Los Angeles Harbor outranked all ports in the world by 1912. The cities around the harbor boomed, too, with Long Beach leading the way, growing in population 690 percent between 1910 and 1914 and emerging as southern California's second largest city. The business generated by the harbor also benefited other sectors, such as finance, construction, transportation, wholesaling, and retailing, and expanded the economy of the entire region. Meanwhile Los Angeles County's population exploded from 15,000 in 1870 to 790,000 in 1914.[27] None of this, however, was natural.

If nature had blessed southern California with certain climatic advantages, it had not been as generous in providing the region with a harbor. San Pedro Bay, a slight indentation on the coastline about twenty miles south of downtown Los Angeles, had been home to various port sites since the eighteenth century. Although Spanish explorer Juan Rodríguez Cabrillo praised the harbor in 1542, San Pedro Bay in reality opened unprotected onto the Pacific Ocean. If San Pedro Bay was a harbor, one settler observed in 1873, it was the largest he had ever seen, for it stretched from the mountains all the way to Japan.[28] The village of

San Pedro, sheltered by the hilly peninsula at the western end of the bay, was the main landing in the early nineteenth century. After a storm destroyed the San Pedro wharf in 1858, Phineas Banning built a pier in an inland lagoon four miles northeast of the village at a site he named Wilmington, after his Delaware hometown. From that more protected location, he launched boats that unloaded cargo ships anchored offshore.[29] Banning's pier was the birth site of an immense harbor district that grew up after the 1880s and today spans more than twenty miles of waterfront from San Pedro to Long Beach and comprises two major ports.

When Banning arrived, however, Wilmington Lagoon lay at the head of a slough in a sandy collection of nearly landlocked marshes, mudflats, and shallow, shifting ocean channels.[30] The only entrance to the lagoon was partly blocked by a sandbar and required ships to negotiate the treacherous Deadman's Island, which had confounded sailors since the first Spanish arrivals. At high tide in the 1880s, the sloughs around Wilmington looked like a "shimmering sea," in the words of one visitor, but when the tide ebbed the town lay "high and dry, on the edge of a long stretch of wet marsh and mud."[31] So shallow were the channels that cattle could cross them to pasture on Rattlesnake Island, the beachhead separating Wilmington Lagoon from the ocean.[32] In short, nothing ensured that the site of Banning's pier would one day be a deepwater harbor. In 1889, as Angelenos lobbied for federal money for harbor development, the visiting chairman of the Senate Commerce Committee chided his escorts from the Los Angeles Chamber of Commerce: "It seems you made a big mistake in the location of your city. You should have put it at some point where a harbor already exists instead of calling on the government to give you what nature refused."[33]

What nature refused, engineering and civic will provided. Banning moved his shipyards several times to escape shoaling in the Wilmington Lagoon, but after the San Gabriel River changed course as a result of floods in 1867–1868, the silt problem greatly diminished. Thereafter, the primary problem facing the port was the shallow and shifting channels, which prevented oceangoing ships from docking at the wharves. In 1871, the U.S. Army Corps of Engineers spent two hundred thousand dollars dredging a ten-foot-deep channel and constructing a breakwater between Deadman's and Rattlesnake islands. In 1881, the main channel was dredged five feet deeper. But in the 1880s, the largest ships still had to be unloaded offshore, and ocean currents continued to drive sand into the channel. Despite the environmental problems, however, by the mid-

1880s, many people believed the young town of Wilmington had "a fair prospect."[34]

In addition to environmental difficulties, the harbor faced substantial political obstacles. Seeking to defray the cost of the breakwaters, jetties, dredging, and other construction efforts necessary to improve the port, the city of Los Angeles turned to the federal government for help. So too, however, did the Southern Pacific Railroad, which owned a pier at Santa Monica Bay, northwest of Wilmington. The railroad company attempted to persuade Congress to locate the harbor there instead. Fearing the dependence on the railroad that the Santa Monica harbor location would bring, the Southern Pacific's many rivals within the Los Angeles business community formed the Free Harbor League to lobby for the San Pedro–Wilmington site. The *Times*'s Harrison Gray Otis, a longtime opponent of the Southern Pacific and very likely a landowner in San Pedro, and Thomas Gibbon, the president of the Terminal Railway Company, whose line ran to San Pedro, conducted a public relations campaign to transform southern Californians' long-standing antipathy for the Southern Pacific's allegedly monopolistic practices into support for the San Pedro–Wilmington site. They also wooed California senator Stephen Mallory White, who likely owed his position on the Senate Commerce Committee to Gibbon's influence.[35] The city leaders, greatly aided by the general public mistrust of the Southern Pacific that simmered in the late nineteenth century, eventually outlobbied the railroad, and Congress appropriated money for the breakwater at San Pedro in 1897.

While Los Angeles Harbor construction was underway, the United States began building the Panama Canal. The canal would cut eight thousand miles off the water route between the east and west coasts and promised to trigger an upsurge in Pacific Coast shipping. By the time the canal opened in 1914, the Army Corps had completely reengineered San Pedro's landscape. The harbor had a 9,250-foot breakwater, 200 feet thick at the base and more than 60 feet high. Ships entered the harbor through an eight-hundred-foot-wide channel that had been dredged to thirty feet deep. More than thirty thousand feet of wharves spanned the former marshes and mudflats. Furthermore, there were now two ports. In 1906, the city of Long Beach had purchased eight hundred acres of tidal wetlands and built its own port with an inner and outer bay, an entrance channel, and three navigational channels. The two harbors were soon to be connected by a two-hundred-foot-wide channel, and plans were afoot to dynamite Deadman's Island for landfill.[36]

From the perspective of the region's commercial elite, harbor growth could not have been better timed. The Army Corps engineer Charles Leeds declared in 1915 that "at the present time when new trade routes are being established due to the opening of the Panama Canal . . . every port on the Pacific is struggling by the preparation of improved harbor facilities, to secure the establishment thereto of these new trade routes." Once established, he feared, trade routes would be difficult to divert. Nothing, he warned, should "be permitted to handicap the harbor of Los Angeles County in the race for port supremacy." The harbor, he proclaimed, was "the greatest single asset of this county."[37]

All of this growth, however, along with the investment made to attain it, made southern California more dependent than ever on the rivers behaving themselves. By 1914, the city of Los Angeles had spent $5.5 million on harbor construction, and the federal government an additional $5.8 million.[38] Private merchants had spent millions more. Despite this investment and the race for Pacific port supremacy, no one took any precautions to protect the county's greatest asset from floods, even though both the Los Angeles and San Gabriel rivers had at times in their capricious histories emptied floodwaters into the harbor site, depositing silt in the tidal wetlands. Since shifting channels during the deluges of 1867–1868, however, the San Gabriel River had delivered most of its waters to Alamitos Bay, several miles east of the harbor.[39] Consequently, as port construction proceeded, less than half of the historical discharge from the rivers emptied into the harbor. During this period, the flood-control question arose twice and both times was disregarded. When the Southern Pacific argued that the Santa Monica location was less flood-prone, which it in fact was, Los Angeles business leaders dismissed the claim as an expression of the railroad's selfish interests, which was also true.[40] Engineers representing the city at an 1896 Army Corps hearing considered the silt problem at Wilmington "a matter of little moment" and insisted that what little silt there was "could easily be diked off."[41]

After the Wilmington site's advocates prevailed, the flood issue resurfaced when the federal Rivers and Harbors Act of 1902 proposed building such a dike. Army engineers, however, blocked the project. The dike would have closed off a slough that the engineers considered a useful waterway for future port expansion. "To construct the dike," the corps feared, "would mean placing a barrier to the future development of Wilmington Harbor." Corps engineers reassured Congress that most of the San Gabriel River no longer flowed into the harbor area and so did not constitute a silting threat anyway.[42]

Thus, the very political rivalries, urban development ambitions, and engineering expertise that created the harbor at San Pedro also derailed early flood-control proposals. Building the city also endangered it. Meanwhile, the Los Angeles River continued to flow into the harbor.

Farming the Flood Plains

At the same time that the harbor emerged from the swamp, plowed fields and orchards replaced the willows and grasses in the valleys and on the coastal plain. As was the case with harbor construction, agricultural expansion meant that humans invested money in rearranging a landscape that was subject to flooding. Recent immigrants like the Germains and natives with short memories plowed and tilled flood-prone lands. As they did so, they benefited from a complex local relationship between water and soil, without which the development of southern California could not have happened.[43] And yet, in the very process of taking advantage of the region's hydrologic blessing, they were also exposing themselves to its flood dangers.

Rain comes to the region in bunches, but subterranean geology evens it out. This is particularly true of the San Gabriel River basin, fifteen miles east of downtown Los Angeles. For hundreds of thousands of years, the stream had coursed out of the mountains, depositing debris on the flatlands of the San Gabriel Valley. In the early twentieth century, the river and its countless tributaries flowed over these alluvial beds, gradually sinking into the porous sediments.[44] Most of the time, the stream disappeared underground almost entirely about six miles downstream from the mouth of the canyon, leaving barely a trickle to flow through the dry riverbeds on the surface.[45] Seeping between the boulders and stones and into the gravel-filled basins below the surface, the water inched its way downhill. It obeyed all the laws of flow of surface water, but it did so much more slowly, so that water entering the basins at the mouth of the canyon took as long as a quarter of a century to make its way a few miles across the valley.[46] Slowly and invisibly, the groundwater headed toward a single outlet east of downtown Los Angeles at Whittier Narrows, where impermeable subterranean formations forced it back to the surface. The valley and the basins below it resembled a giant bathtub: water flowed in at a northerly faucet at the mouth of San Gabriel Canyon and drained out the southern end at the Narrows. The enormity of the underground basins and the inexorable flow of groundwater stored years of runoff at a time. In effect, the basins evened out the

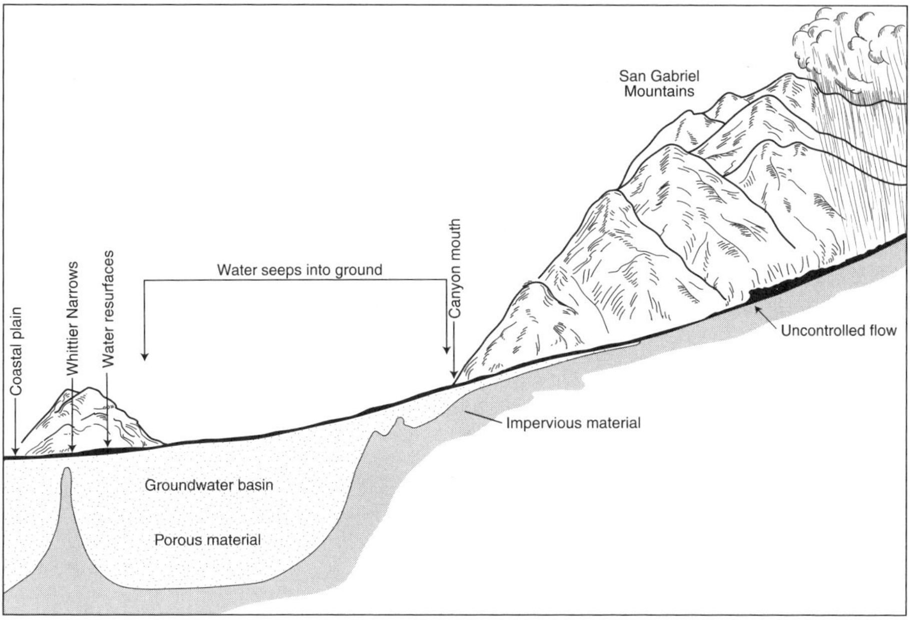

Map 3. San Gabriel Valley groundwater basin cross-section. Prior to urbanization, rivers leaving the mountains seeped into the porous soils of the San Gabriel Valley and flowed through gravelly groundwater basins toward Whittier Narrows, where impermeable subterranean geological formations forced them to resurface. (Map by Carol Marander.)

peaks and dips in the river's discharge, so that groundwater remained relatively constant, fluctuating annually by only a small amount in comparison to the enormous swings in input.[47]

San Gabriel Basin hydrology not only supplied water for agriculture; it also fertilized the soil. Centuries of floods depositing sediments had made the riparian lands the richest in the region. Furthermore, the river channels' frequent shifts had delivered silt even to lands that by the 1880s were several miles from the nearest waterway. Not surprisingly, then, much of the agricultural expansion took place in floodplains. The floods of 1889–1890, for example, filled a slough south of the Los Angeles city limits. The swamp dried up in 1894, and the stench of rotting fish wafted over the surrounding area, prompting farmers to burn and bury the carcasses. The fish fertilizer enriched the soils of the now-dry slough, which was cultivated with alfalfa and beets during the next decade. It was "good ground," one farmer later recalled.[48] Similarly, terrain along the San Gabriel River came under plow. Riparian lands valued at forty dollars an acre in 1894 sold for many times that in 1914.[49] Some land prices soared as high as a thousand dollars an acre.[50] North of the town of Whittier, where a tangle of existing and former San Gabriel channels crisscrossed, so much cultivation covered the maze of waterways that engineers surveying the river in 1913 could barely discern the old channels.[51] Thus, many of the lands that came under cultivation for the first time after 1880 had been flooded in the past and could be again. The best places to farm were also the most flood prone.

The effectiveness with which nature converted storm runoff into groundwater storage and distributed fertile soil across the plain fueled the first commercial agriculture booms.[52] Because of the scant rainfall and surface water, the perishability of farm products, and southern California's isolation from other markets, the region's agriculture prior to the 1870s served local consumers almost exclusively, the main exception being animal products that were part of an international trade, such as hide and tallow. Fruits, vegetables, and grains were produced only on a scale sufficient to meet local needs. Starting in the 1870s and 1880s, however, when farmers first tapped the groundwaters, and as transcontinental railroads opened new markets for Los Angeles produce, commercial farming boomed.

The most dramatic gains came in the citrus industry. In 1862, some twenty-five thousand citrus trees bore fruit for local and statewide markets. In 1877, the year after the transcontinental railroad arrived in Los Angeles, growers sent the first boxcarload to the East. By 1881, there were five hundred thousand trees in the county, and citriculture was just getting started. During the next three decades, the improvement of refrigerated rail cars, the hybridization of species, and the organization of elaborate management and advertising techniques created a thriving citriculture industry. By 1914, there were more than one and a half million orange trees in the county, and some thirty thousand rail cars of fruit, valued at more than twenty million dollars, went eastward each year.[53] With underground basins watering the crops and railroads carrying them east, a new landscape emerged where cattle had grazed fifty years before. Citrus orchards checkered the base of the San Gabriel Mountains and the vicinity of the Narrows. Walnut trees and small vegetable farms covered the valley floor, and dairying, alfalfa, and sugar beets predominated on the plain.[54]

The impact of commercial agriculture, however, had far-reaching economic effects. Like harbor construction, the citrus industry was integral to the region's urban boom, with oranges and other fruits constituting extremely profitable export commodities that generated capital for regional investment and created linkages to other sectors of the economy. The elaborate technological, economic, and transportation arrangements that were necessary for the industry stimulated infrastructure development from which the entire region benefited, and they encouraged cooperation among growers, who formed organizations to try to rationalize prices, labor relations, crop and pest management, and advertising. Moreover, thousands of smaller growers with rising disposable incomes supplied the consumer demand that fed the initial bursts of manufacturing and retail in the region. As the citrus industry boomed, so did the rest of southern California.[55]

Thus, groundwater and the rearrangement of nature and society to take advantage of it nourished the young metropolis. In the San Gabriel Valley alone, the number of irrigated acres doubled in the first decade of the century, representing not only an increase in the total amount of land under plow but also a rise in the value of that land and in the investment in irrigation, transportation, marketing, and technology required to make the land profitable. The total value of farm property in Los Angeles County jumped from $74.8 million in 1900 to $200 million in 1910.[56] At the same time, the cities along the rivers grew too. Between 1890 and 1920, the population of Whittier grew from 1,926 to 12,531. The population of a number of other towns overlying the underground basins grew from one- to sixfold.[57] With farm and city supplied by the enormous groundwater reserves, Los Angeles and its vicinity blended the pastoral with the urban.

Both harbor construction and commercial agriculture, then, transformed the landscape, and both represented substantial investment in rearranging nature. Both were cornerstones of the region's bustling economy and its prospects for future growth. By 1914, the city builders had ventured much on the presumption that it would not rain.

FLOOD PLAINS UNDER PAVEMENT AND PLOW

If the past was any indication of the future, however, it surely would rain again in southern California. Although most people dismissed the region's record of flooding as the events of exceptional years, in some senses it was they who were living in exceptional years, for nature's historical violence had helped create the environment that the people found so welcoming. Over a period of time unfathomable to the newcomers at the end of the nineteenth century, flooding, channel changes, dry spells, coastal tides, and unstable mountain geology had sculpted a terrain and nurtured vegetation well-suited to the region's wild variations in precipitation. In this ecosystem, deluges on occasion radically altered the face of the landscape but in general did little long-term damage. This ecology persisted into the late nineteenth century, when urban physical features—which humans believed they were erecting with nature's blessing—altered the landscape to make flooding not only more likely but also more damaging. In other words, the harbor, agriculture, and the cities, all of which southern Californians were building because of the environment's temporary benignity, were speeding the end of nature's benevolence.

The floods had made the land. For millennia the waters of the Los Angeles and San Gabriel rivers had gathered sediments in the mountains and cascaded onto the plains. There, the residue of thousands of years of floods lay on the vast detritus cones that fanned out of the canyons along the mountain front for fifty miles or more. On these fans, the rivers began to drop their load, the heaviest chunks first, the finer material farther downstream. As the riverbeds' gradients fell, the waters spread over the valley floors, leaving fertile beds of alluvial sediment.[58] Through repeated floods, the silt piled up, barricading the water, which eventually broke free to find a new channel. The fertile ground left behind gave birth to vegetation. Settlers in the nineteenth century described the coastal plain as lush and marshy mosaics of grasses, willows, alders, and cottonwoods. When the water overflowed these thickets, the vegetation slowed it and spread it, causing it to drop its silt. The soils that settlers in the late nineteenth century found so fertile were a product of the very climatic action they did not believe in—repeated deluges and constantly changing river channels. As one old-timer observed, "It is foolish to say that it hurts the land to have it overflowed, for that is what made it."[59]

But a few decades of cultivation and urbanization undid what many years of flooding had accomplished. During storms as late as the 1880s, the willow thickets along the river channels had slowed the torrents, often preventing them from eroding or overflowing the banks, and harmlessly spreading the waters that did inundate cultivated lands.[60] The development of the next few decades, however, altered the ecology so that smaller floods would cause more damage. The harbor is a good case in point. In many ways, the harbor created its own flood problem. As the engineer Charles Leeds noted in 1915, Los Angeles was very fortunate to have begun its harbor construction when silting was at a low ebb. Between the absence of major floods and the change in the San Gabriel River, which historically had emptied into the sloughs that later became the harbor but now temporarily flowed into Alamitos Bay, the port area received a comparatively low silt load during the years that construction was being planned and initiated. Furthermore, prior to construction of the harbor, nature had made its own arrangements for removing the silt. The wetlands on which the harbor was eventually built were a transition zone between the land and the sea. Slowed by vegetation and shallow, winding channels, rivers dropped their silt in the flat marshlands. The tidal action of the ocean, which lapped against the wetlands, scoured away the silt. The development of the late nineteenth century disrupted this bargain between the rivers and the sea. The construction of the breakwater calmed the water adjacent to the mouths of the rivers. Stilling the waters was, of course, the main goal of harbor improvement, but it had the unintended effect of creating a giant settling basin for the debris the rivers discharged. Without the tidal action to scour away even the little debris that the rivers brought to the ocean during the dry early years of the twentieth century, the harbor required constant dredging to maintain depths sufficient for large ships to enter the port.[61] The sudden and massive siltation that would come with a major flood could potentially bring shipping to a halt. Thus the development of the harbor, the heart of the region's growing commerce, aggravated its own flood problem, a problem that concerned people so little at the time of port construction that they took no precautions to avert it. The harbor is but one example of urbanization after the 1880s altering the hydrology of southern California to make the region more flood prone.

Railroads, pavement, and plows were also responsible for this ecological change. Embankments for railroads and bridges that traversed the region's river channels often trapped debris from overflows, causing water to seek new channels (which often had farms and houses in their paths by this time). Also, railroad or bridge embankments in a stream could block passage of water through one portion of the bed—causing the same amount of water to be squeezed through a smaller opening, thereby increasing the velocity (and the erosive power) of the torrents. The surveyor S. B. Reeve later blamed such alterations made by railroads for most of the destruction caused by the 1914 floods.[62]

Less visible, but even more transformative in the long run, was the spread of impermeable surfaces that came with urbanization. Water overflowing sand, gravel, marshes, and woodlands soaks into the ground. Rooftops, roads, and storm drains, in contrast, inhibit absorption by the ground and instead concentrate the water and channel it into faster streams. On a typical square-mile of land that was subdivided into two-and-a-half-acre plots, roads covered about 28 percent of the total surface area. Between 1904 and 1914, the city of Los Angeles alone gained nearly five hundred miles of improved roads. These roads were mostly macadam, a composition of small stones held together with asphalt and oils, which was almost completely impervious to water. After 1898, southern California began spraying the roads with asphalt, and by 1915, the city of Los Angeles had paved nearly all its streets.[63] F. A. Coffman, who lived near the San Gabriel River, complained in 1915 that these new features of the landscape "catch the water and shoot it down to us in a short time." He warned that "this is something to be watched and accounted for, otherwise it will cause much damage."[64] In 1915, Captain Charles Leeds, who had overseen the Army Corps's harbor construction, confirmed Coffman's observations: "By the growth of cities and towns, with their great areas of roofs and paved streets, by the extension of paved highways . . . the run-off resulting from any given rainfall is steadily increasing." He called it a matter of "extreme importance."[65] Thus began a problem that would escalate the flood menace for the rest of the twentieth century: as runoff concentrated more quickly and in larger amounts, it had greater potential for eroding, overflowing, or smashing whatever lay in its path.

The expansion of impermeable surfaces proved particularly damaging in combination with the increase in land brought under plow. Agriculture removed native vegetation and loosened soil, enabling overflows to erode the land more. In the mid-nineteenth century, when Jose Ruiz's father had settled in southern California, only four or five ranchers occupied the land south of Los Angeles between the city and San Pedro. So thick did the blackberries, willows, and tules grow, Ruiz recalled, that a rider on horseback could scarcely get through in some places. After the 1880s, however, farmers began to clear the brush. They burned the trees for firewood, drained the marshes, and sank wells that lowered the water table, killing the remaining vegetation.[66] They replaced the thickets and grasslands with crops. By the 1910s, most of the land between the harbor and the downtown was under plow, relegating the remaining stands of willows and grasses to riverbeds and riparian lands.[67] This development would later worry county engineers. "Whereas in the earlier history of this county flood waters could spread harmlessly over wide areas whose soil was bound down by grass and other vegetation," they warned the Board of Supervisors in 1915, "the soil flowed over is now that of well-cultivated orchards and beet fields and hence is easily eroded, only to be deposited at some point nearer the ocean." Furthermore, they were concerned that water recently imported by aqueduct from the Owens Valley, three hundred miles away, would increase the runoff flowing into the Los Angeles River. In the future, they predicted, "uncontrolled run-off will do far more damage through erosion and silting than has been done in the past." The increasing runoff, the engineers forecasted, "will change moderate floods of the past into serious floods in the future."[68]

Ecologically, Los Angeles was very different from and much more hazardous than the place it had been in 1884. Roads and buildings covered a substantial and increasing portion of the land. Farms with loosened soil covered most of the rest. Bogs became croplands. Sloughs turned into harbors. And railroads crisscrossed riverbeds. Willows and grasses disappeared. By 1914 it took less rain to make a flood.

UNPRECEDENTED RAIN

Shortly after midnight on 18 February 1914, rain began to fall on southern California.[69] It continued for sixty-six of the next seventy-nine hours. Heavy rainfall in late January had already soaked the ground, leaving it unable to absorb more runoff. Water rose in the riverbeds and in the streets of the towns. In residential areas north of Los Angeles, the rising rivers threatened homes. Verdugo Creek, a tributary of the Los Angeles River, undermined a two-story house that had sat 150 feet from the stream. In five minutes the house caved in, disappearing into the roaring currents. Meanwhile wealthy citizens in the Arroyo Seco district scrambled to protect their homes. As debris backed up against the bridge embankment on Avenue Forty-three, water gushed into the neighborhood. A hundred men assembled rock barriers to reinforce the canyon walls, but when the water rose again on the evening of the twentieth, the river ripped their work apart in minutes and resumed its assault on the community. At eleven o'clock that night, a house on the corner of Homer and Avenue Forty-three burst into flames, illuminating the devastation in the black night. As residents scrambled to escape the inundated neighborhood, twenty homes collapsed into the billowing waters. Eventually the city dynamited the bridge to free the waters and direct them back into the main channel.[70]

Through city streets the water raged, unable to find a place to soak in or settle. The "fine broad pavements" of the city of Los Angeles, the *Examiner* reported, "were totally blotted out by long and broad stretches of turbulent water." Moving south of Los Angeles, the water overflowed fields and orchards, sweeping away crops and drowning livestock. The *Times* described the scene south of town as a "wide rushing lake."[71] Picking up soil from the plowed fields, the torrents plunged into the port, choking it with more than three million cubic yards of silt, discoloring the harbor all the way out to the breakwater. Where water had previously stood to a depth of twenty-four feet at the municipal wharf, mounds of mud now poked above the surface.[72] Two steamers attempting to navigate the channels became mired in the mud, unable to move for days.[73] By the time the sun peeked through the clouds around seven o'clock on the morning of the twenty-second, the deluge had done more than ten million dollars worth of damage, including a price tag of four hundred thousand dollars for dredging the county's greatest asset.[74]

The damage caught southern California by surprise. The *Times* editorialized that people of the area had never "experienced such a flood" and, indeed, had never received any "indication that a flood of such force was possible." The mayor said the deluge "was not to be anticipated." The *Examiner* called it an "unprecedented rain."[75]

These and other declarations, however, contradicted much other evidence. The dozens of old-timers whom engineers interviewed in the aftermath of the 1914 disaster almost unanimously agreed that the floods of 1862, 1867–1868, 1884, and 1889 were larger in terms of the land area they inundated and the total volume of water they discharged.[76] Engineers, who calculated the maximum discharge of floods of the 1880s at 60 to 70 percent greater than that of 1914, confirmed the old-timers' recollections.[77] Another engineer estimated the extent of the land that the 1914 flood overflowed to be only one-fourth to one-third that of the 1884 and 1889 deluges.[78] The *Times* was simply wrong; Los Angeles *had* flooded before.

In other ways, however, the *Times* was right. Los Angeles, the infant metropolis, had never experienced anything like the deluge of 1914. Since the storms that the old-timers recalled, much had changed. In the 1880s, southern Californians believed in the essential benignity of nature, but they had not yet come to rely on that presumption. For the most part, they considered the spreading waters harmless, which indeed they often were. Southern Californians had not yet built an eleven-million-dollar harbor, nor had they plowed up the flood plain. The city was not yet filled with newcomers ignorant of past overflows and lulled into inaction by the dry climatic phase. Willow thickets still prevented erosion. Permeable soils still absorbed overflows. And tides still scoured the shoreline. In short, in the 1880s, southern Californians did not yet depend on the absence of floods, nor had they altered their environment in ways that made it more flood prone. By 1914, in contrast, a city had emerged on the plain, and with it a new ecosystem. Even though the ecological changes remained invisible to the sun-drenched populace until 1914, it was a much more hazardous ecosystem. Consequently, in the context of the February storm, the new and seemingly unrelated facets of urban life, each harmless by itself, came together with destructive force unprecedented in the region's previous experience with flooding. Los Angeles, however, claimed to have learned its lesson. As the *Times* opined in the days after the 1914 flood, "Experience is a reliable teacher."[79] Never again would southern Californians treat their rivers with such nonchalance.

CHAPTER 2

A Centralized Authority and a Comprehensive Plan

Response to the Floods, 1914–1917

The flood impelled southern Californians to try to control water. In July 1914, 250 representatives from municipalities, civic organizations, and businesses gathered downtown at Blanchard Hall in response to an invitation from the Los Angeles County Board of Supervisors. One delegate calculated the average interval between floods over the previous century and warned that deluges "will cause greater damage in future years." He concluded that something had to be done: "Supine indifference and lack of energy to meet these flood damages is not in harmony with the successful achievement of this County." To preserve that successful achievement, the delegates resolved that "control of the flood waters of the County should be vested in a centralized authority and should be in accordance with a comprehensive plan embracing the entire county."[1]

The resolution signaled a break with earlier responses to flooding. It replaced the exceptional-years thesis with an estimation of the average interval between floods, and it exchanged faith in a benign nature for forecasts of future calamity. Also, flood control previously had been piecemeal, the domain of individual cities, landowners, and small flood-protection districts, but now the delegates called for a "centralized authority" and a "comprehensive plan." No longer would they tolerate the "exceptional years" that brought torrents in some seasons and barely a trickle in others. No longer would they tolerate the haphazard and ad hoc flood-control efforts that left only some parts of the county protected, often at the expense of others. Instead, they envisioned, engineers would redesign the rivers, the public would finance the plans, and nature would submit to human management. In attempting to execute these visions, the delegates institutionalized a new ideal of hydraulic order.[2]

That institutionalization occurred through two simultaneous processes. One was the formulation of a plan for controlling the waters. Here southern Californians drew upon Progressive Era faith in entrusting social problems to objective experts. So great was their mistrust of the rivers after the 1914 flood and so great was their confidence that experts could tame the waters that they dispatched a team of engineers even before settling on who would pay for the resulting plan or by what authority it would be implemented. Consequently, another part of institutionalizing flood control entailed the establishment of a government body with the authority to execute whatever plan the engineers produced. In this second task, southern Californians encountered a fragmented political terrain: neither state nor federal law provided any agency with the power to conduct comprehensive regional flood control, and disagreements between the city of Los Angeles and the rest of the county threatened to defeat the entire effort. The resulting flood-control regime reflected both the impulse to control water and the conflicts that arose as southern Californians crafted a new public authority. Thus, a new flood-control regime was born when the waters of February 1914 mixed with the culture of the Progressive Era and the political terrain of turn-of-the-century California.

THE COMPREHENSIVE PLAN

The rain of February 1914 inspired much heroism. As water ten feet deep swirled around the bedridden Jeanette Williams, the patrolman William Rice dove into the billows. Bystanders cheered as he emerged from her home minutes later, swimming with the ailing woman on his back.[3] Even H. H. Rose, the mayor of Los Angeles, rushed into the rain to help people to higher ground in the Arroyo Seco neighborhood.[4] The next morning, Mayor Rose addressed the citizenry from his bed, nursing the cold he had caught in the Arroyo. Last night, he declared, the torrents rendered the citizens "practically helpless," but next time, he promised, Los Angeles would be ready. He pledged to direct city engineers to "devise a means of safeguarding . . . districts affected by the heavy rains."[5] From every corner of the county, people echoed his resolve, inundating officials with pleas for flood control. The following week, the County Board of Supervisors called a public meeting, at which it resolved to appoint a team of engineers to produce a flood-control plan. On its editorial page, the *Los Angeles Times* printed a cartoon in which an engineer, labeled "Los Angeles," bent over a table with pencil and ruler in hand. Plans for numerous civic projects, including a bundle of scrolls marked "Plans for Repairing Storm Damage," cluttered his office.[6] In the next flood, southern Californians resolved, human ingenuity and a coordinated preventative strategy would obviate the need for the heroics of Rice, Rose, and others. As he switched from rescuing people in the rain to preaching about prevention, Mayor Rose embodied a larger shift in flood control. Local flood lore had always lauded individuals who displayed bravery during emergencies, but the new flood-control heroes would be engineers.

The new engineering mania was not an inevitable or purely technical response to flooding, but rather sprang from historical circumstance.[7] In contrast to the late nineteenth century, when everything in southern California's ecology and society worked to convince people that flooding was unusual and benign, the 1914 flood and the Progressive Era historical context in which it happened combined to persuade southern Californians that they faced a severe hazard that only concerted human effort would alleviate. The deluge washed away the exceptional-years thesis and shocked people into a recognition that urban growth required comprehensive control of the rivers. Meanwhile, the Progressive Era zeal for rationalizing everything from factories to forests provided a new lens through which the flood-control convention delegates viewed the disruption to their metropolitan environment as they began formulating their comprehensive plan.

The calamity of February 1914 revolutionized how people thought about floods. If southern Californians had previously treated inundation as a colorful exception to the normally benign behavior of the environment, they decided in 1914 that floods were frequent and dangerous.[8] In March, the Los Angeles County Board of Supervisors appointed a team of engineers to study the deluge. The engineers read old court cases and Army Corps documents to study the region's flood history. They also interviewed the historian James Guinn, as well as dozens of old-timers, who recounted the great historic overflows.[9] The research convinced the engineers that flooding was a regular event. "Judging the future by the past," they reported to the supervisors on 3 June, "a repetition of disastrous flood may come in two years." Compiling a list of all the past floods they could find, the team calculated that Los Angeles should expect a damaging flood every 3.25 years.[10] "The Imminence of the flood danger, and the ever increasing cost of right of way demand the earliest possible action," the engineers recommended in their final report the following year.[11] The engineers, like many of the old-timers they interviewed, were convinced more damaging floods would follow in the future if the county did nothing to prevent them. Nineteen-fourteen was not an exceptional year.

The floods also taught southern Californians that the region's environment was not completely benign. The silt at the municipal wharf and the waterlogged houses in the Arroyo signaled that the city builders' metropolitan dreams could not coexist with the unruly rivers.[12] The great ancient civilizations, one of the county's engineers speculated, had collapsed because of their inability to tame nature's fury. Los Angeles must undertake comprehensive flood control, he urged, in order to "insure against the same fate which overtook these ancient peoples." We must, he recommended, "Divide and Conquer," in "attacking these great floods . . . before they have become united and irresistible."[13] Another engineer warned that "Nature" does not give "any warning to man, nor does she show any respect for his ready made channels."[14] And Mayor Rose condemned the "savage waters" that "were not to be denied their prey."[15] In 1914, as savage waters stalked their prey and humans schemed to divide and conquer, a struggle between nature and society was emerging in people's minds. It seemed that either nature would flood the metropolis every 3.25 years, costing millions of dollars and disrupting trade and communication, or people would tame the rivers, confining them to official channels, and forcing them to flow in an orderly manner.

If the experience of the flood alerted people to the problem, progressivism told them what to do about it. Rose and his compatriots were not the first southern Californians in history to fall prey to the region's savage waters, but they were the first to view that experience through the lens of progressivism. Roughly spanning the period from the late 1890s into the 1920s, the Progressive Era was a time when, nationwide, societal problems, especially those concerning the environment, were coming to be seen as ever larger and more complex, requiring more elaborate organizations for combating them.[16] People formed private organizations and government agencies to pool their resources to attack problems that they were unable to solve as individuals. Most frequently, they turned these problems over to experts. With objective expert management, everything in society, it seemed, from U.S. Forest Service chief Gifford Pinchot's national forests to automobile mogul Henry Ford's assembly lines, could be made to function systematically and efficiently. It was this obsession with system and expertise during the Progressive Era that linked people of different professions, social backgrounds, and ideological bents. Through a progressive lens, the sludge in the Los Angeles Harbor, the disrupted business, and the unforeseen costs of repairs did not appear to be exceptional events in a usually benevolent natural environment. Instead, they were disruptions to a system. Order had broken down.

In appointing the team of engineers in March 1914, the Board of Supervisors hoped to restore that order. Four of the five members of this Board of Engineers Flood Control were prominent local civil engineers with extensive backgrounds in water management. Henry Hawgood, the chair of the engineering board, had worked for the Southern Pacific Railroad on water problems and had called for a comprehensive county flood-control program as early as the 1890s. Frank Olmsted had reported to the supervisors on San Gabriel River flood problems in 1913. Charles T. Leeds had served as district engineer for the Army Corps of Engineers, Los Angeles District, and had overseen part of the construction of the Los Angeles Harbor. J. B. Lippincott had won renown for his efforts to secure water for Los Angeles from Owens Valley. The board's fifth member, James W. Reagan, was less well known, a newcomer to the region after having worked as an engineer building railroads in Kansas, Arizona, Mexico, and South Africa. Each of the first four engineers took responsibility for a geographical portion of the watershed and focused his efforts on devising a plan for controlling floods in that section. Reagan undertook the task of determining the area that previous deluges had historically overflowed. Although they would work mostly autonomously, the five planned to meet regularly to discuss their findings and to review each other's assessments. They intended in the end to combine their studies into a single report for the supervisors.[17] This division of labor would have disruptive consequences down the road.

Over the next year, the engineers picked the river system apart and studied it piece by piece. First they dispatched parties to gather field data. They surveyed stream bed gradients, searched the landscape for high-water marks, projected the channel width necessary to contain flows, calculated how to desynchronize the flood peaks, measured surface water absorption rates, and assayed harbor silt to determine where it was coming from. Having disassembled the rivers and analyzed their component parts, the engineers reported to the supervisors in July 1915 on how to redesign the system to bring order to each segment. For the mountains, the engineers advised reforestation to minimize runoff and small check dams to lessen the amount of water flowing at the peaks of floods. They also advocated several medium-sized masonry dams to catch water and release it slowly. In the foothills, they planned to spread excess waters over the porous debris cones, reducing overflow and recharging groundwater supplies for irrigation. On the coastal plain, the engineers envisioned speeding the water between levees of straightened official channels, reinforced with brush, rocks, or wooden pilings. Finally, they proposed to protect the harbor by diverting the Los Angeles River so that it would empty into the ocean east of the port.[18] Overall, the team promised to regulate the flow of water from the mountains to the sea. They would dam it, spread it, confine it, and divert it until the rivers functioned with the same efficiency and mechanical predictability that Henry Ford had achieved on his assembly lines and that Gifford Pinchot had extracted from his forests.

Despite the precision with which the engineers expected to manage the water, disorder and dissension plagued their attempts to do so. From the beginning, Reagan marched to the beat of a different drummer. While his colleagues pored over blueprints and scoured remote canyons searching for high-water marks, Reagan drove twenty-five thousand miles around the county, meeting people and listening to their concerns. On the basis of these excursions, he and his assistants compiled six hundred pages of interviews with old-timers who shared their memories of great floods past and their opinions on what to do now. These interviews convinced Reagan that flood control was needed where the people were—downstream. He had no use for the mountain check dams and erosion control that the other engineers advocated. The cost of these works in relationship to the property and population they protected, he insisted, was too high, and to rely on checking the waters in the mountains and storing them on the gravel cones would, he said, turn out "disastrously" for the lower district. On the basis of what his informants told him about their downstream needs, he mapped what he believed to be the most menaced zones of the county and, usurping the responsibility of the other engineers, outlined his own flood-control program for the entire county, emphasizing downstream flood-control works. Reagan's unorthodox approach strained his relationship with the other board members, and he even stopped attending their meetings.[19] Tempers flared at one Board of Engineers meeting in August 1915 when Reagan assailed his colleagues' flood-control proposal. "If you weren't satisfied with the work," Hawgood demanded, "why didn't you meet with us and convince us of our errors?" "That would have been too big a job," Reagan retorted.[20] Eventually Reagan submitted a separate final report to the Board of Supervisors, which he did not share with his colleagues for review beforehand.

More than a year of engineering studies by the Board of Engineers had failed to yield a single best plan. Instead, there were two best plans. The majority report suggested types of strategies rather than specific projects. Reagan's version offered more specifics and in that regard was closer to what the supervisors had asked for. His plan, however, lacked the backing of the four distinguished experts that the majority report enjoyed. The presumption behind the Flood Control Association's call for a comprehensive flood-control plan had been that a single, objectively best solution existed and that experts could discern it. The conflict among the engineers indicated otherwise. In the end, the supervisors decided to accept both studies, designating Reagan's the minority report.[21]

Despite their technical differences, the two reports together outlined a vision of a new hydraulic order. Rejecting the exceptional-years thesis and drawing on Progressive Era sensibilities, the volumes' authors suggested that the rivers were naturally ornery but that engineering could tame them and make them run like machines calibrated to serve society. The thornier question was how to implement the plans.

THE CENTRALIZED AUTHORITY

Engineers were not the only ones who disagreed among themselves. During the year the Board of Engineers spent surveying and calculating, public officials and other prominent southern Californians cooperated to craft an institutional structure for executing the forthcoming comprehensive plan. By threatening the future of the metropolis, the 1914 deluge had initially united southern Californians, but that momentary accord veiled deep divisions. Although they began with confidence that they could tackle their problems "as one," flood-control advocates quickly discovered that their consensus receded with the floodwaters of 1914.[22]

Although the flood-control advocates deployed the Progressive Era language of objective expertise, the theory that policy making could be streamlined by objective expert management proved to be dramatically out of step with flood-control events between 1915 and 1917. Neither federal nor state law authorized any existing agency to carry out such comprehensive flood control as southern Californians had in mind. As flood-control advocates began constructing such an entity, public consensus on what constituted the common good fractured. By 1917, their efforts yielded a grotesque mockery of Progressive Era faith: a flood-control program directed by Reagan, the unorthodox and controversial engineer, who was in charge of implementing a plan that he personally opposed and that the public only barely approved. Thus, the Los Angeles County Flood Control District, the bureaucratic bulwark the flood-control advocates created to impede the floodwaters, contained within it a contradiction between the Progressive Era affinity for objective, expert policy making and the legacy of discord from which the agency was born.

While the engineers worked out their proposal, business leaders and municipal officials explored possibilities for executing that plan. The flood-control advocates immediately rejected the practice of entrusting oversight to the small protection districts. The work done by these districts, according to the Long Beach Chamber of Commerce, was "woefully inadequate and was wrecked by the water it was supposed to control. . . . [It was] damaged beyond repair and is useless."[23] Instead of dividing power among many small, impotent districts, southern Californians sought, in the words of one landowner, a "benevolent despotism" to centralize flood-control authority.[24] The most desirable source of that despotism, however, was unclear. Divisions of power among federal, state, and county governments left none of them in charge of problems such as the one southern Californians now faced. Consequently, when flood-control advocates decided to create a new authority, that decision was in large part a product of the fragmented political structure in turn-of-the-century California.

By the time the supervisors dispatched the engineers into the field in 1914, California had a long tradition of establishing weak state institutions and vesting more authority in local governments.[25] In the realm of flood control, this pattern resulted largely from nineteenth-century struggles over the rivers of the Sacramento Valley. Most frequently, the conflicts pitted advocates of a strong centralized government authority to coordinate a valleywide flood-control plan against others who favored leaving flood control in the hands of private landowners or small, localized districts.[26] Localism had prevailed throughout the second half of the nineteenth century, and by 1914, state law delegated flood-control responsibility to local, quasi-public flood-protection and drainage districts such as those that had formed in southern California in the 1890s and 1900s. These districts were composed of landowners who petitioned county boards of supervisors for the right to tax themselves and build their own bulwarks against the rivers. There was no statewide authority to coordinate or regulate the activities of these districts, and they often worked at odds with one another.

Bitter intrastate regional rivalries further fragmented state authority. Throughout the nineteenth and early twentieth centuries, mutual suspicion among regions, especially between northern and southern California, prevented locales from securing state funding for many types of projects, from hospitals to irrigation. These rivalries were so fierce that they periodically inspired movements to split the north and south into separate states.[27] More frequently, however, the conflict resulted in state government stalemate and the delegation of responsibility to local authorities. In 1880, for example, southerners opposed the legislature's efforts to pass the Drainage Act, which created a state flood-control program to check the hydraulic mining debris choking the rivers of the Sacramento Valley. The *Los Angeles Herald* declared that it "would be just as logical to tax the whole State to pay for a failure of the crops and fleece which have been ruined by drought in the southern counties as to levy a tax to repair the ravages of the debris of mines."[28] After the measure passed, southerners joined with other regions of the state that objected to being taxed for the Sacramento Valley's benefit and challenged the law in the courts. In 1881, the state supreme court declared the act unconstitutional, eliminating state coordination of flood control and restoring the anarchy that allowed each local district to devise its own protection.[29] Chronically underrepresented in the legislature, southern California in particular suffered the effects of such conflicts.[30] In 1914, southern Californians believed (correctly, as it turned out) that state funding of some sort would be forthcoming eventually.[31] Given, however, the long and contentious history that had fragmented state government and fueled interregional rivalries, the state was not a reliable authority in which to vest power for taming the rivers of Los Angeles.

Nor was there much federal help on the horizon. After the 1914 flood, county officials and private citizens alike called for federal flood-control assistance on the grounds that protecting the Los Angeles Harbor from silt was a navigation problem. As a precedent, they pointed to the federal California Debris Commission, which had investigated, though not implemented, a flood-control plan for the Sacramento River in the first decade of the twentieth century. This federal involvement on the Sacramento, however, had emerged under somewhat fluky conditions. In 1884, a federal court resolved the hydraulic mining problem in California by outlawing the practice, in part because mining tailings were impeding navigation of the Sacramento River and San Francisco Bay. At the urging of miners, however, Congress passed the Caminetti Act of 1893, which created the California Debris Commission, which was composed of three officers from the U.S. Army Corps of Engineers and charged with reestablishing and carefully regulating hydraulic mining in the state. A little-noted phrase in the act empowered the commission to take steps toward "affording relief from flood." The Army Corps ignored the apparently inadvertent clause for a decade, but after a flood in 1907, northern California farming and business interests urged the commission to take up its flood-control powers under the act.[32] Thus, the federal involvement on the Sacramento River was an unintended by-product of a mining law and was not seen as particularly applicable to Los Angeles.

Moreover, although the Army Corps expressed tentative interest in Los Angeles flood control in 1914, until the New Deal, the corps's flood-control work was narrowly restricted to promoting navigation. Given the seasonal character of most southern California waterways, the navigation requirement effectively limited federal participation to the small portion of the floodplain immediately affecting the harbor. There was a likely possibility that the corps would help build a channel to divert the Los Angeles River around the harbor, but the mountain dams and valley levees proposed by the Board of Engineers lay beyond federal jurisdiction at the time. With Sacramento abdicating authority over flood control and offering only stingy appropriations for local projects, and with Washington construing its responsibility to apply only to navigation, Los Angeles flood-control advocates decided they could not "wait for the [federal] government nor for the State."[33]

The only remaining source of central authority was the county. Although the *Times* called it "the organization best fitted to get maximum results in a minimum of time," even the county government was an imperfect agent for combating the deluges.[34] As both the executive and legislative power in the county, the Board of Supervisors had the authority to establish additional special protection districts if it wanted. These districts were easy to create and could make policy with a simple majority vote of landowners within their boundaries, but they lacked the crucial power of eminent domain necessary to carry out the right-of-way acquisitions the comprehensive plan would require. As Olmsted reported to the supervisors, the current law did not "offer scope enough for the work now under contemplation." And while the Board of Supervisors could exercise such authority, the county itself would be an unwieldy instrument for flood control, because financing such projects required a two-thirds majority vote from the citizenry.[35]

Flood control advocates decided to negotiate this fragmented political terrain by state approval to create a new local governmental structure, a metropolitan-wide flood-control district that would combine the simple majority rule of the small districts with the eminent-domain powers of the county. The vehicle they created to establish this new authority was the Los Angeles County Flood Control Association. The association was a quasi-public organization whose purpose was to assist the Board of Supervisors in drafting legislation to create a new flood-control authority and to generate public support for the undertaking. Unofficially, the Flood Control Association formed just after the 1914 flood, when the Board of Supervisors invited southern California civic leaders to a mass flood-control meeting in February, and it formalized its rules and membership at a subsequent gathering the following July. In all, more than one hundred organizations joined the Flood Control Association. City governments could each send three representatives, as could most civic organizations, such as chambers of commerce and boards of trade. Railroad companies were allowed one representative. The organization made policy with simple majority votes of its members.[36]

The Flood Control Association delegates had much basis for consensus. They did not differ much socially or ideologically. From the Long Beach Chamber of Commerce to the Los Angeles Realty Board to the Southern Pacific Company, the membership of the Flood Control Association was homogeneous, composed almost exclusively of business leaders and landowners. The membership ranged politically from the conservative Supervisor R. W. Pridham, who chaired the association, to progressives such as Supervisor John J. Hamilton.[37] There were no labor organizations or dissident socialists involved. All members fancied themselves civic leaders who could look out for the interests of the public because they themselves had a stake in the orderly future development of the metropolis.[38] One railroad official, for example, said, "Some may say we are prejudiced by the fact that our rails follow the river for some miles. Perhaps we are. . . . But as taxpayers and dependents on the prosperity of Los Angeles and vicinity, we wish to see something done quickly which will prevent romping floods."[39] Thus, the association was composed of a batch of elite civic leaders who shared an interest in flood control and an inability to accomplish it on their own.

Their common interests, however, did not prevent acrimony. As the Flood Control Association delegates set about fashioning a metropolitan flood-control district, the deepest divisions were geographical. Delegates from the city of Los Angeles favored a plan to finance the district's work with a special assessment on the properties that would benefit most directly from flood-control projects. The city of Los Angeles was by far the largest city in southern California, home to more than half of the county's population. Parts of the downtown had historically flooded, but the bulk of the city's population—including many of its wealthiest and most influential citizens—lived west of the Los Angeles and San Gabriel watersheds, for the most part out of harm's way. Under this scheme, the city would reap indirect benefits from the improved agricultural and commercial conditions that flood control would provide, without having to shoulder much of the cost. Instead, that burden would fall on the riparian landowners outside the city boundaries. In January 1915, the Los Angeles County deputy counsel Charles Haas submitted to the state legislature a bill to create such a special-assessment district.

Delegates from outside the city of Los Angeles protested. Their objections reflected more general growing pains of southern California municipal government after the turn of the century, when many problems were getting too large to be financed by special assessments on individual landowners. Moreover, benefits of projects such as roads, sewers, harbors, and utilities diffused over the entire population, making the value of those benefits difficult to calculate and leading many landowners to conclude that levying special assessments only on affected property owners was burdensome and unjust. In the first two decades of the century, therefore, the financing for municipal public works projects increasingly came from the taxes on the entire population. These dissident Flood Control Association delegates insisted the same logic should apply to flood control. Most of them represented areas along the San Gabriel and lower Los Angeles rivers, sparsely populated territory dominated by large riparian landowners who would face the largest assessments under the Haas plan. To avoid this tax burden, the delegates retained Glendale attorney Frederick Baker to craft an alternative bill, and they threatened to enlist northern California legislators to oppose the Haas special-assessment measure. Baker's bill planned to finance the district by levying a *districtwide* property tax that would spread the costs of flood control to all property owners within its boundaries.[40]

Tension filled the downtown meeting hall where the Flood Control Association gathered on the last day of March 1915 to debate the two versions. A. W. Fry, a banker from Clearwater, outside the city of Los Angeles, championed the Baker bill, arguing that flood-control benefits were too broad to be financed by local special assessments. One of his opponents rejoined, "[The fact that] Mr. Fry's land lies along a river and threatens to be washed into the Pacific Ocean every time a fog comes up is no reason why we should stand the cost of anchoring it for him."[41] Ultimately, Fry and his allies, outnumbering the Los Angeles delegates nearly two to one, forced the Flood Control Association to approve the Baker bill. In the sixty-two to thirty-three vote, the city delegates lost to those from the rest of the county. Fearing that such internal division would jeopardize the bill's fate in the legislature, the delegates then revoted on the motion to make the decision unanimous in order to convince the legislature of the "pressing importance" of flood control.[42]

Meanwhile, the Haas special-assessment bill continued working its way through the legislature. When both bills passed, the bewildered California governor, Hiram Johnson, asked the supervisors which one he should approve. On their recommendation and against his better judgment, he signed the Baker general-tax bill in June 1915, a month before the Board of Supervisors received the two reports of the Board of Engineers. Johnson, like the dissenting delegates from the city of Los Angeles, put aside his particular objections in the name of the urgency of getting any sort of flood control underway.[43]

Out of this discord emerged the centralized authority, the Los Angeles County Flood Control District. Under the provisions of the act, the new district encompassed all of the county except the offshore Channel Islands and the desert area north of the San Gabriel Mountains, and the Board of Supervisors served as the district's executive body, though the new agency was technically a separate entity from the county. These arrangements sidestepped the state constitutional provision requiring a two-thirds majority vote of the citizens to pass county bonds. The district would be headed by a chief engineer, appointed by the Board of Supervisors, and would be staffed by the number of employees the engineer deemed appropriate. They would not be subject to civil-service regulations, a point that would later cause great conflict. The chief would oversee the development of flood-control plans and submit them to the supervisors, who would then put the recommendations on election ballots for the public to approve. If the public endorsed the plans, the supervisors could then sell bonds to raise money for projects, which would be financed by the controversial districtwide property tax. With few exceptions, once the bonds were approved projects could not be changed.

Thus the public could ratify or reject the experts' designs, but it had little ability to affect the formulation of those plans.[44] By transferring so much decision-making authority to engineers, the framers intended to prevent special interests, of which Progressive Era people were so suspicious, from tampering with the objective expert plans and hoped to scuttle the types of political disputes that had plagued the creation of the district.

The discord, however, only intensified as the Flood Control District started its work. The Board of Supervisors met on 30 August 1915, to consider its next move. The meeting began tensely, as the members addressed each other with frosty parliamentary politeness. As the testy supervisors descended into innuendo about flood-control "improprieties" and speculation about one member's rumored impending retirement, tempers flared. They interrupted each other and accused each other of digressing from the topic at hand. "Haven't we had enough discussion?" Chairman Pridham finally snapped.[45] At that point, they began taking a series of votes that eventually led to the board selecting the controversial James Reagan as chief engineer of the new agency. Each of the votes split three to two, with the supervisors representing areas outside the city of Los Angeles defeating those who served constituencies within. The rancor into which the meeting descended was an extension of the conflict that marked the work of the Board of Engineers and the Flood Control Association. Like the engineers and the delegates, the supervisors agreed on the need for centralized flood-control engineering, but they disagreed on how costs and power would be shared. These controversies revolved around one man.

James Williams Reagan was a maverick. After studying civil engineering and a variety of other subjects at a college in Kansas, he took a job teaching Latin. He left after a year and bounced between engineering jobs in Mexico, South Africa, and the United States before settling in Los Angeles in 1907. During this time when engineering was growing increasingly professionalized, he learned most of his trade on the job. Later, he quit the American Society of Civil Engineers under clouded circumstances, thus sacrificing the most important mark of professional standing.[46] He had further set himself apart in the summer of 1915, when he refused to sign the majority report of the Board of Engineers and instead submitted his own minority recommendations. His supporters compared him to Abraham Lincoln, the self-schooled statesman, and to William Mulholland, the self-taught aqueduct builder. Reagan "is not a book engineer," Supervisor Hamilton said in his defense. "He is an outdoor engineer."[47] His critics, however, derided his slim credentials and charged that he was "incompetent to design the simplest kind of an engineering structure."[48]

What Reagan lacked in professional standing, however, he made up for in political savvy. One of his Southern Pacific Railroad employers described him as "exceedingly diplomatic and successful in handling the public."[49] And one of his successors at the Flood Control District remembered him as "a rather astute political thinker" who "wouldn't hesitate to pull any strings he could politically."[50] He was good at charming newspaper editors, politicians, and the public, exhibiting this skill during his excursions around the county interviewing people about their flood memories and cultivating a constituency.

In doing so, he got caught in the simmering conflict between the city of Los Angeles and the rest of the county. As an extension of their quarrel over who should pay for flood control, city and county residents also disputed which works to build. Riparian property owners with whom Reagan had spoken insisted that flood protection required what came to be known as downstream works—the reinforcing, enlarging, and straightening of channels on the valley floors. The length and number of these waterways, however, and the excavation and other work needed to protect them promised to make the downstream channel works very costly. Los Angeles civic leaders, in contrast, preferred instead the less expensive upstream check dams and reforestation that the Board of Engineers' majority report had recommended. The only item the two sides agreed upon was the need to divert the Los Angeles River from the harbor, and even here, they disagreed on the route.[51]

Possibly in part because of his ties to the Southern Pacific Railroad, a major landowner on the coastal plain, Reagan had adopted the downstream advocates' view in his minority report to the Board of Supervisors. Deriding the check dam proposals, he declared that downstream channel work was "absolutely imperative as an initial engineering procedure," a first step in a multifaceted plan that included both upstream and downstream works. Anything else would be "unfair and unjust, and should not and could not receive the support of the people."[52] In the weeks before the supervisors' heated meeting, citizens in the San Gabriel Valley and lower Los Angeles River basin had showered the supervisors with petitions urging them to name Reagan as chief engineer.[53] As the board debated Reagan's appointment, Supervisor Hamilton argued, "The citizens who live along these rivers have met him and realize that he is in sympathy with them. He values what they know, and he has organized the knowledge of citizens along these streams into a comprehensive plan of what ought to be done."[54] After Reagan's appointment, however, the Los Angeles Chamber of Commerce and other organizations protested that "the value of his plans [is] at least open to question" and urged that the Flood Control District save money by adding flood control to the duties of the county surveyor instead of appointing a new officer.[55] Although Reagan, Hamilton, and the Chamber of Commerce all spoke in terms of engineering expertise and public welfare, the outcome boiled down to power politics: Reagan's constituents had three representatives on the Board of Supervisors, while his opponents controlled two seats. Amid all the talk about engineering imposing order on nature, conflicting interests were injecting disorder into engineering.

Ironically, the author of the minority report had now been appointed to carry out the recommendations of the majority report. As Reagan proceeded in 1916, however, it became apparent that he was not following exactly the recommendations of the now disbanded Board of Engineers.[56] The plan he completed the following year called for a multifaceted attack on the floods. It included three mountain dams, numerous check dams in the canyons, and a channel to divert the Los Angeles River around the eastern side of the harbor. The plan relied most heavily, however, on downstream work, or "river training," as Reagan called it. The beds would be straightened and reinforced with wooden pilings, wires, and brush to confine the streams to designated channels. This downstream emphasis provoked vehement opposition from the city of Los Angeles, the Municipal League, the *Los Angeles Times*, and others. As the Board of Supervisors deliberated whether to put Reagan's recommendations on the ballot, his two enemy supervisors tried to divide the ballot measure and let citizens vote on the more universally popular harbor-diversion plans separately from the rest of the proposals in hopes that the more divisive aspects of the program would go down to defeat. The Board of Supervisors, however, as was becoming its habit with flood-control issues, again split three to two in favor of putting Reagan's whole plan before the voters as a single $4.45 million bond measure. As voters went to the polls on 20 February 1917, the chasm that divided the city from the rest of the county on this issue manifested itself in the election returns. Although the city of Los Angeles voted it down, the measure carried by extraordinary margins elsewhere in the county. The final tally showed it passing by a slim majority. It might not have passed at all without the severe flooding a year earlier in January 1916, which had reminded voters of the threat they faced from their rivers.[57]

The fractured flood-control consensus, however, was not initially a problem for the young district. Later that year, the United States entered World War I, during which the federal government suspended the sale of municipal bonds, thus delaying flood-control construction until late 1918. Then, in 1919, the Flood Control District received a no-conditions-attached $250,000 grant from the state of California. The legislature increased that amount to $300,000 in 1921, a figure it would renew annually until 1933, when it discontinued appropriations to the district.[58] By 1920, the Flood Control District and its contractors had managed to line dozens of miles of streams with temporary wood and wire protections and to construct Devil's Gate Dam to tame the Arroyo Seco, a seasonal stream that ran through the hills between Pasadena and Los Angeles. Most important, the district cooperated with the Army Corps of Engineers to build the channel diverting the Los Angeles River around the harbor. Modest as these steps were, they represented a fundamental break from the period prior to the 1914 flood, when, as Reagan said, even "the idea of attempting to curb and regulate the flood waters . . . was new with the people."[59]

PUBLIC WORKS, ASSEMBLY-LINE STYLE

By 1917, southern Californians had institutionalized a flood-control regime. With the Board of Engineers' recommendations as a "comprehensive plan" and Reagan's Flood Control District as a "centralized authority," both the idea and the structure that would shape flood control for the rest of the century were in place. The idea was that nature was a disorderly but knowable system, which humans could redesign using technology to cause rivers to function in orderly, predictable ways. The structure organized engineers in government bureaucracies to discern and implement the single best strategy for redesigning the rivers. For the next two decades, the Flood Control District's personnel would study water and design projects under the direction of the chief engineer, who would recommend engineering strategies to the Board of Supervisors. That body would put the proposals before the people. Those that the people ratified would become ironclad flood-control policies, except when something went very wrong.

Thus institutionalized, Los Angeles flood control mirrored trends in American political culture. The Progressive Era witnessed an organizational revolution that increasingly located policy making in labyrinthine bureaucracies that managed public and corporate activities in the twentieth century. Three characteristics distinguished this new policy-making style. First, it vested much power in unelected experts who drew authority from their appointments in governments and later in universities. Second, policy usually grew out of an alliance between governmental bodies and private economic interests, an alliance that justified its claims to represent the public interest by citing the opinions of the allegedly impartial experts it retained to investigate public problems. Finally, although the public was not literally shut out of the decision-making process, policy debates, which so frequently revolved around technical issues and excluded political and moral ones, inhibited participation by nonexpert citizens. After developing this technocratic style of decision making in numerous small laboratories such as Los Angeles flood control, Americans experimented with it on a large scale during World War I, as government and industry planners managed the wartime economy and domestic mobilization. With the Great Depression and World War II, technocracy emerged as the dominant American policy-making paradigm, and by the middle of the twentieth century it shaped policy making on issues from nuclear power to economic planning to national defense.[60]

From the beginning, institutionalized flood control in Los Angeles followed this national pattern. Experts on the Board of Engineers counseled the government officials, business leaders, and landowners in the Flood Control Association, and together they established a flood-control district that allowed citizens to ratify or reject the technical decisions of engineers but otherwise to have very little input. Even in the 1930s, when the district's role was substantially redefined as the U.S. Army Corps of Engineers assumed most of the region's flood-control planning, the basic arrangements of expert engineers crunching hydrologic data into designs for physical structures to direct the flow of water and to protect economic investment in Los Angeles would translate from local to federal management without a hitch.

There was, then, a curious parallel between the flood-control advocates' visions of how both rivers and flood-control districts should work. These advocates envisioned that water would flow through rivers and policy would move through bureaucracies in a controlled, predictable, perpetual manner, producing desirable outputs at the end—step by step, as though on an assembly line. But river flows without variations and policy making without politics were contrary to how both rivers and policy were wont to run. If any step in the operation deviated from plan—if natural phenomena or the behavior of flood-control structures turned out to be unpredictable, if engineers disagreed on the best designs, or if voters approved projects and then changed their minds—the whole system might sputter to a halt.

By 1917, plenty of irregularities already hinted at that possibility. The inability of the Board of Engineers to come up with a unitary plan, the fragile flood-control consensus of 1915, and the maverick chief engineer all clouded the future smooth functioning of the Flood Control District's operations. Contrary to engineering wisdom and Progressive Era culture, both of which held that through rational design nature and society could be made to function smoothly, these small sources of instability would blow up in the 1920s and 1930s into substantial problems. Neither nature nor politics nor engineering would behave as an orderly component of the urban ecosystem. Instead, they would interact in unforeseen ways, first to delay flood-control construction in the 1920s, then to produce a new type of flood disaster in the 1930s, and finally to bring an end to locally led flood control.

CHAPTER 3

A Weir to Do Man's Bidding

The Great San Gabriel Dam Fiasco, 1917–1929

In its 1915 majority report, the Board of Engineers estimated that it would take only five years and $16.5 million to construct a network of check dams, diversion channels, and other devices to tame the county's waters. These structures, they predicted, would "permanently relieve the people of Los Angeles county from the menace of future floods."[1] By the end of 1929, however, all the Flood Control District had to show for its efforts were a few successful projects, one enormous failure, a lot of wasted money, and a total loss of credibility.

It was a dam that derailed the flood-control assembly line in the 1920s. Departing from the modest recommendations of the Board of Engineers, the Flood Control District concentrated its energies during the 1920s on building a twenty-five-million-dollar barrier, 425 feet high, across San Gabriel Canyon in the mountains northeast of Los Angeles. The San Gabriel Dam, which at the time promised to be the world's tallest, would smooth out an uneven landscape. It would slow the waters that gushed out of the mountains and release them under control into the valleys. It would also eliminate the cycles of drought and flood by storing water from the wet years for use in the dry. Mountain or plain, rain or shine, water would flow in a regular manner. "It will stand," the *Los Angeles Times* boasted, "as a giant concrete weir to do man's bidding and say on the one side to the raging flood waters of the Sierra Madres, that they may come so far and no farther and, on the other side, that the fertile valleys to the sea may be cultivated in security from either drought or torrent."[2]

But the dam's proponents encountered a political terrain just as uneven and treacherous as the physical landscape they sought to reengineer. First, drought, urban growth, and politics combined in the early 1920s to launch the quixotic project, even though it departed substantially from the Board of Engineers' 1915 comprehensive plan. Almost immediately, however, disagreement arose over how to implement the proposal, and construction was delayed for five years. When the project finally got underway, unstable rocks revealed both the faulty geology beneath the dam and the political corruption behind it. Not only was the dam itself unbuildable, but the scandal it engendered undermined the Flood Control District's capacity to control the flow of water elsewhere in the county. As it turned out, neither the weir nor the technocratic policy-making style would do Los Angeles's bidding.

THE GREATEST DAM IN THE WORLD

On 6 May 1924, Los Angeles County voters went to the polls to decide the future of what some called the "greatest dam in the world."[3] The thirty-five-million-dollar bond proposal on the ballot that day called for the construction of eleven large dams in the mountains, including a gargantuan one in San Gabriel Canyon. The *Los Angeles Times,* which had opposed the flood-control bonds in 1917, urged its readers to approve this project.[4] Voters, the *Times* said, could choose between floods that "tear down across the fertile lands" or "properly controlled" waters that would leave "prosperity in their pathway."[5] The voters overwhelmingly chose prosperity. They approved the measure by the largest majority any bond measure had received in county history.

Between 1917, when the electorate had only barely passed the $4.5 million bond measure, and 1924, when thousands followed the *Times*'s example and reversed their flood-control stance, much changed. The Flood Control District's chief engineer, James Reagan, bucked conventional engineering wisdom and added big water-conservation dams to the district's agenda. Southern California suffered several years of drought. And a real-estate boom enriched the county and knit the interests of city and countryside together. By 1924, these developments convinced one and all that building the world's tallest dam in San Gabriel Canyon offered the best way to defend against both deluge and drought.

Historical circumstance, just as much as rational engineering, was guiding flood-control policy.

It is doubtful whether Los Angeles could have mustered the economic resources or the political will to undertake the San Gabriel Dam project in 1917. The Flood Control District's 1917 plan, a less expensive program of much smaller scale, enjoyed the support of only a slim majority of the public. Moreover, engineers considered such a structure economically unfeasible. San Gabriel Canyon was too steep, they said, to afford good dam sites, and the rapid erosion rates would give any such structure a short life. In a 1913 report on the San Gabriel River, the engineer Frank Olmsted contended that the only structurally feasible site would require a 215-foot dam that would cost more than it would yield in flood-control benefits. The 1915 majority report of the Board of Engineers called for only three large dams in the mountains but no concrete dams and no large ones of any sort in San Gabriel Canyon. In that report, Olmsted reiterated his earlier finding: there were no good dam sites on the San Gabriel River.[6]

Not surprisingly, one of the first engineers to challenge this assessment was James Reagan. Although he generally agreed that a mountain dam would be hard to justify economically in 1917, Reagan also believed that the growing metropolis at some point might be able to afford such a project. In the late 1910s, Reagan felt there were two problems facing the county. The most pressing task was immediate protection for the downstream lands. This problem, he was confident, could be met by the small dams, channel reinforcements, and harbor diversion projects already underway.[7] Even after the 1917 bond projects commenced, however, another problem remained. "Nature," Reagan said, "has a very intermittent, spasmodic and wasteful way of recharging our underground rivers."[8] He therefore also wanted to link flood control to water conservation. The next step, he believed, should be to control the water in the mountains, "at its source." This vision for mountain stream control, however, was far more ambitious than the system of check dams that the Board of Engineers had recommended in 1915. Reagan wanted to build a structure so large that neither the worst imaginable storm nor the longest drought could threaten the region. He wanted a dam in the mountains, a big one.[9]

At some point in the late 1910s, he began to investigate such a possibility in his spare time. Weekends and holidays and late into the nights he worked, charting the groundwater basins and studying logs of wells in the San Gabriel Valley. He sent surveyors into the canyon looking for dam sites, eventually settling on Twin Forks, where two branches of the San Gabriel River came together. A 425-foot structure there would impound 240,000 acre-feet of water—enough, he believed, to store the largest flood ever recorded on the San Gabriel.[10] The dam he imagined would stand higher than Idaho's Arrowrock Dam, then the world's tallest, and would cost twenty-five million dollars, a fivefold increase over the 1917 bond issue, the projects of which were not yet complete, and half again more costly than what the Board of Engineers had envisioned spending on the entire flood-control program. Even more significantly, the project promised to shift the goals of the Flood Control District's work, giving increasing importance to another great Progressive Era cause, water conservation. In suggesting such an enormous dam, which promised to change the cost, scale, and purpose of flood control, the maverick chief engineer charted a radical departure in the district's program.

As Reagan was planning to overhaul the flood-control program, Los Angeles kept growing. In the late 1910s and early 1920s, oil, motion pictures, and real-estate development sparked a boom that was remarkable even in fast-growing southern California. Meanwhile, as midwestern migrants followed newly completed transcontinental highways to Los Angeles in droves, new urban and suburban communities sprang up in previously rural areas of the county, and the populations of older cities such as Long Beach and Pasadena also grew at phenomenal rates. In less than four years, three hundred thousand residential lots were subdivided, forming a nearly continuous mosaic of development across southern California and increasing intraregional interdependence. By annexing the San Fernando Valley and other rural parts of the county, the city of Los Angeles more than quadrupled its size between 1912 and 1929 and also increased its own stake in the well-being of neighboring agricultural areas.[11] As Reagan put it, "The entire County seems destined to be one great city."[12]

The future development of that city, a joint committee of the region's chambers of commerce said in 1923, depended "upon an adequate and controlled water supply." Water projects, the committee maintained, benefited both urban and agricultural zones by increasing "taxation and general prosperity."[13] By the early 1920s, Los Angeles was not only in greater need of conservation of its floodwaters but also wealthier and more potentially unified in that need. Under such conditions, one engineer declared in 1920, a dam in San Gabriel Canyon was "feasible both from the standpoint of engineering as well as economy."[14]

In 1920, Reagan invited the U.S. Bureau of Reclamation's renowned engineer, Arthur Powell Davis, to consult with him on the flood works underway. In an era of great dams, Davis was the greatest dam builder. He had supervised the construction of Arrowrock Dam as well as of several other of the bureau's grandest monuments.[15] His opinion counted. Reagan excitedly took him out to the Forks to tell him about his vision. "What a huge thing I was going to show him," Reagan recalled. Davis was impressed. "This is the best place in the basin for the dam," he said. "It is very big. Where are you going to get the money to build it?"[16] Under the provisions of the 1915 Flood Control Act, Reagan had to get it from the voters.

To gain the citizens' approval, he laid the foundations for a carefully orchestrated dam campaign even before most voters had heard of San Gabriel Dam. In February 1923, at the dedication of the newly completed San Dimas Dam, Reagan spoke to a crowd of five hundred and asked the citizens not to abandon their commitment to the flood-control program.[17] That fall, he led representatives from the county's chambers of commerce on a tour of the Flood Control District's construction sites. Among the stops on the weekend-long sojourn was the proposed San Gabriel Dam site. Upon their return, the delegates promptly petitioned the Board of Supervisors to allocate fifty thousand dollars for Reagan to continue studying the project.[18] When the board complied, the project thought impossible in 1917 was now at least conceivable.

If urbanization helped make the dam economically feasible, drought helped make it politically possible. In the early 1920s, southern California suffered one of its worst dry spells ever. Rainfall was down between 1917 and 1926. The season of 1923–1924 was the driest year since the U.S. Geological Survey had initiated stream flow recordings in 1892. The San Fernando Valley had to cut back agricultural use of the imported Owens Valley water to maintain an adequate supply for the cities. Meanwhile, elsewhere in the county, wells went dry and groundwater levels lowered. The artesian well area had shrunk from 177 square miles to only 28 square miles between 1890 and 1923. Alarms sounded. Reagan reported to the Board of Supervisors in February 1924 that the water supply would be insufficient in less than five years.[19] The joint committee of chambers of commerce warned that a "shortage of water . . . is imminent."[20] In retrospect, the shortage was more imagined than real, but the dread persisted nevertheless. "We are living in a country of uncertain rainfall," the *Times* cautioned. "Past experience shows that we may be deluged with rain or subjected to drought."[21]

Against the backdrop of water scarcity, a huge dam on the San Gabriel—big enough to store water for irrigation and still have space to catch a flood—seemed a welcome remedy. On 1 April 1924, Reagan offered plans for a twenty-five-million-dollar dam in San Gabriel Canyon as well as for ten other dams with a total cost of ten million dollars more. Reagan planned for water to flow continuously through San Gabriel Dam, but only in quantities necessary to recharge groundwater. Any extra water would remain behind the dam until engineers could release it under control. The flow of the capricious San Gabriel would thus be carefully managed to lessen the flood peaks, even out the seasons, and eliminate the effects of the wet and dry cycles. Not a drop of water would flow to the ocean.[22] "By this method," Reagan told the Board of Supervisors, "it is hoped that [water] conservation will entirely take the place of flood control."[23] Against nature's unpredictability, Reagan offered the orderliness of engineering. The board distributed twenty thousand copies of Reagan's report to citizens and placed its proposals on the ballot as a thirty-five-million-dollar bond measure for voters to consider on 6 May.

In contrast to 1917, when flood-control factions had squabbled over how to spend a mere $4.5 million, the 1924 measure garnered widespread support. During a whirlwind campaign, Reagan stumped the county, pitching his plan as a water conservation measure. "The time has come," he told a meeting of the Chamber of Commerce in Glendale, "for Los Angeles County to stop talking about flood control and to undertake a program of conservation of the waters that are allowed to run to waste every year."[24] He wrote to the Board of Supervisors that the burden of water shortages "would not fall alone upon the rural or ranch country, but would find its way quickly in through the financial institutions of the county." Los Angeles must have the dam, he said, "or decline will inevitably follow." San Gabriel Dam, he promised, would forestall this apocalypse and instead make "a perennial garden spot of the entire area."[25] In the San Gabriel and San Fernando valleys and the Long Beach area, all of which stood to gain both flood protection and a more reliable water supply, nearly every community supported the measure.[26]

Old foes of the 1917 bonds such as the *Times* and the Los Angeles Chamber of Commerce got behind the measure too.[27] As the *Times* observed, "This has been a dry year and the county would do well to have a battery of dams filled with reserve water."[28] With the dam, the newspaper said, "floods would be leashed to aid man's needs."[29] Perhaps most appealing to the old flood-control foes in the city of Los Angeles, the dam promised to eliminate the need for the expensive downstream works that had since 1917 absorbed the Flood Control District's energies and budget without providing many direct benefits to the city.[30]

Thus Reagan, the *Times,* and other supporters presented the question in sharp contrasts: did voters prefer inevitable decline or a perennial garden spot—dry years or reserved water? Given such choices, citizens passed the bonds overwhelmingly. Ninety-nine percent of the precincts approved it, and the total vote yielded a three-to-one majority, with some communities returning 98 percent approval rates.[31]

On the morning after the election, the *Times* called the balloting a "tribute to James W. Reagan."[32] Reagan had indeed been an indefatigable promoter, but the passage was more a reflection of the new flood-control politics that water scarcity and urbanization had brought about. In 1915, the Board of Engineers had judged a dam in San Gabriel Canyon economically infeasible, and voters in 1917 only barely approved a $4.5 million flood-control program that many people considered too expensive. Yet less than ten years later, the electorate returned a record majority for a thirty-five-million-dollar program, including a mammoth dam in San Gabriel Canyon. In the interim, urban growth had made the county wealthier and water more valuable, thus enabling southern Californians to spend more to control its flow. At the same time, a drought raised the specter of water shortages, presenting a stark contrast between the uncontrolled river and the orderly flow that San Gabriel Dam promised. Seven years earlier, the dam had been economically and politically infeasible. With drought and urbanization, however, *not building* the dam now appeared infeasible. The capricious climate and growth-addicted political economy of the city, which had together created the flood problem, were now apparently combining to chart its resolution.

COMPETING VISIONS OF HYDRAULIC ORDER

"New wheels of industry began to rumble," the *Times* reported in May 1924, as Reagan and his engineers set about "harnessing the great torrents."[33] Those wheels, however, soon ground to a halt. Within days after the election, engineers began disputing the best design for San Gabriel Dam. Then the project got tangled in a water-rights competition that fissured the public consensus of 1924. Finally, a municipal election and a series of court cases reorganized both the project and the district

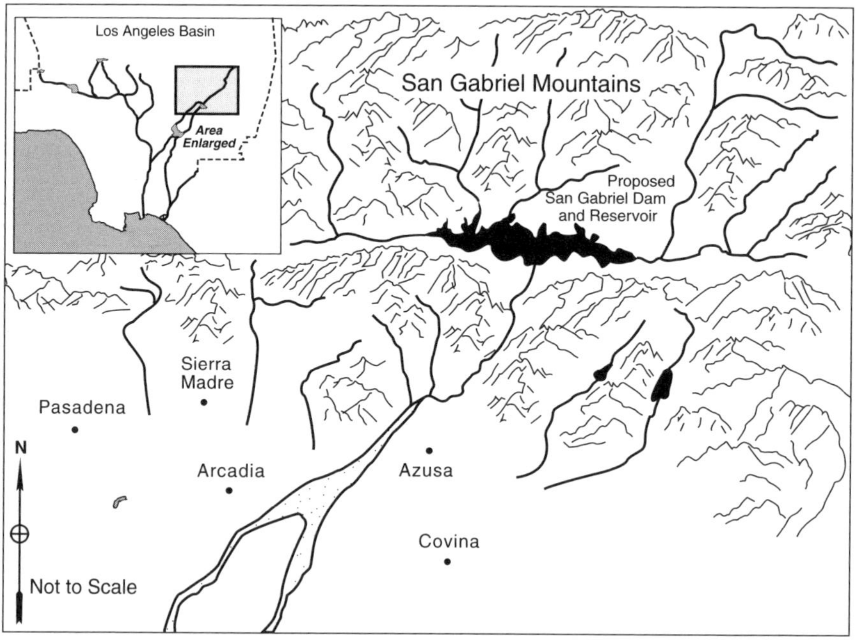

Map 4. Proposed site of San Gabriel Dam, circa 1920s. See also figure 6 in this book's photo section. (Map by Chris Phelps.)

itself. At the end of four years of policy chaos, southern Californians were not a single step closer to being able to direct the flow of water out of San Gabriel Canyon. As the district sought to replace nature's "intermittent, spasmodic and wasteful" way of moving water, it stumbled over its own intermittent, spasmodic, and wasteful way of assembling flood-control policy.[34]

At first, the San Gabriel Dam project seemed to model orderly flood-control policy making. As dictated by the 1915 Flood Control Act, the chief engineer had studied the canyon and filed a report with the supervisors. They had adopted his findings and put them on the ballot, and voters had given the measure overwhelming approval. In 1924 and 1925, however, a parade of notable civil engineers whom Reagan invited to consult on the project began to voice concerns. It was standard civil engineering procedure for directors of public works projects to request second opinions from independent consultants, but the results of these reports shook public confidence in San Gabriel Dam. In a report dated one day before the 1924 election but not publicized until afterward, the Bureau of Reclamation's Davis suggested that a smaller dam "would be a much better investment." In October 1924, the Panama Canal builder General George Goethals pronounced the available hydraulic data "too meager" to determine the necessary reservoir capacity. The authors of an additional consulting report, John Galloway and D. C. Henny, approved the site but refused to certify that it was the best one.[35] Although Reagan steadfastly denied that any of these were significant criticisms, it was publicly evident that the experts did not all agree with Reagan.

The critique with the most impact came from Harvey Hincks, a civil engineer from Pasadena, who suggested an alternative site for the dam in 1925. In contrast to Davis, Goethals, and Galloway and Henny, who worked with the Flood Control District as independent consultants, Hincks was employed by the district's staunch enemy, the Municipal League of Los Angeles. An organization of civic-minded business leaders, the league fancied itself the voice of the people. Its publication, the *Municipal League of Los Angeles Bulletin,* frequently referred to the league and its allies with terms such as "disinterested civic organizations," and it often claimed to be acting "for the taxpayers."[36] Since its founding in 1901, the Municipal League had backed city-planning measures and public ownership of utilities and had tirelessly advocated wise spending by municipal government.[37] The league's leadership cringed when in 1924 Reagan proposed to shift the entire thrust of the flood-control program with a twenty-five-million-dollar plan for a project that some experts considered unbuildable. Worse yet, in the league's eyes, Reagan asked the citizens to approve all this on only a month's notice. Finally, the league distrusted the Flood Control District's exemption from civil service requirements. The league did not consider any of this either wise spending or careful planning. Without success, the league pressed the supervisors to delay the vote for at least six months to make "an adequate check on the plans proposed, and to inform the public sufficiently."[38] For the rest of the decade, the Municipal League continued to oppose the San Gabriel Dam, which it believed violated every principle of orderly government.

After poring over an earlier study by another Municipal League consultant, Hincks proposed an alternative site that he said offered more dam for less money. A dam at Granite Dike, six miles downstream of Reagan's proposed Forks location, Hincks contended, would create a reservoir with less surface area and thus a lower evaporation rate. It would require less concrete because the canyon was narrower. And it would better control the river because a greater portion of the watershed lay above it. Best of all, it would cost ten million dollars less than Reagan's plan.[39] Maintaining that Hincks's Granite Dike plan would be superior to Reagan's improvement of the river, the Municipal League publicized his report in 1925 and demanded that the Board of Supervisors appoint independent engineers to investigate the alternative. The league's vision of order in San Gabriel Canyon involved a smaller, more cost-efficient dam at a different location.

The supervisors refused. Their own experts told them the Dike site was physically and legally infeasible. That site, according to Reagan, would actually store less water than the Forks site, and the chief engineer further dismissed Hincks's plan as a safety hazard. Granite Dike lay at the intersection of two earthquake faults, he said, and no suitable foundation for a dam existed there. Further investigation of the site, he maintained, would waste public money.[40]

Even more persuasive to the supervisors was the advice of the county lawyers. The Flood Control District's counsel reminded the board that section 15 of the Flood Control Act bound the district to construct the projects "in conformity with the report, plans, specifications and map" that the voters had approved, unless—and here was the phrase that would generate so much controversy—"some change of conditions" made the project illegal or undesirable.[41] In the event of such a change, a four-fifths majority vote by the supervisors was required to approve alterations to the plans. Hincks's study, the lawyers advised, did not constitute a sufficient change of conditions to warrant deviation from the voter-backed plans. Even to spend public money to investigate the Dike site, they said, was illegal.[42] Thus the board's experts disagreed with the Municipal League's experts on what the public interest was and how it should be executed. And by requiring the demonstration of changed conditions and a supermajority vote by the supervisors before any project could be modified, the structure of flood-control law virtually foreclosed the possibility of altering—or even debating—a project once the bonds were passed. Under these circumstances the supervisors had little room to resolve the controversy.

When this disagreement entwined with a water-rights struggle, the unresolved conflict spread to the next step in the flood-control assembly line, stalling the technocratic machine. The Flood Control District needed a right-of-way across the Angeles National Forest in order to build a rail line to supply the San Gabriel Dam construction site. The city of Pasadena also coveted the right-of-way for a water-storage reservoir it planned to build elsewhere in San Gabriel Canyon. After both the district and the city of Pasadena filed for a right-of-way in 1925, the federal government formed the San Gabriel River Commission, a three-member panel of engineers from the Interior and Agriculture departments (the agencies that administered the land in question) to adjudicate the dispute. Commencing in February 1926, the commission's hearings pitted two publics, the Dike supporters and the Forks advocates, against each other. Pasadena favored the Dike dam, because it would not interfere with the city's own reservoir ambitions. San Gabriel Valley residents, who stood to benefit the most from both the flood protection and the water conservation that the Forks dam would provide, turned out to support the Flood Control District's petition. Further intensifying the dispute, Pasadena was located outside the San Gabriel Basin, which meant two things: first, the city had little to gain in the way of flood protection from the San Gabriel Dam, wherever it was built; and second, the valley residents viewed the city as an interloper trying to gain rights to water that they believed rightfully belonged to them. Thus flood control got entangled in a bitter water-rights dispute. The two sides represented distinct, irreconcilable publics.

The testimony before the federal San Gabriel River Commission in early 1926 was a model of technocratic debate. Each side marshaled a litany of engineers to plead its case, and Reagan submitted a fifty-three-page technical report reviewing dam specifications and hydrologic data for the commission and accusing Pasadena of basing its claims "upon assumption" instead of on "actual and positive facts."[43] Amidst this technical discussion, both sides competed to appear more public spirited. In his opening statement, Pasadena's lawyer offered a variety of compromises that, he assured the commissioners, "would serve the purposes of all the main contending parties." We can, he added, "protect every right of the users in the valley."[44] The Forks site's proponents, meanwhile, condemned Pasadena for blocking the people's will. "The Republic has failed in functioning," they charged. "The people are not supreme. . . . We voted for what we wanted on May 6, 1924, but we did not get it. . . . The will of the electorate has been nullified, the popular mandate has been defeated."[45] Both sides found that the language of expertise and public interest was sufficiently flexible to support their claims. With neither the experts nor the public able to agree on what the public interest was, much less the best method for executing it, the technocratic pursuit of a single most-efficient solution broke down. As one Forks site proponent put it, too many "obstructions . . . [had] been thrown into the machinery of the Flood Control Department."[46] While the federal commission deliberated, construction of the world's greatest dam was on hold for yet another year.

Meanwhile, the conflict spread into municipal politics, turning the 1926 local election into a referendum on Reagan and San Gabriel Dam. Southern California was growing faster than the flood-control program could provide protection to the region. Fearing the flood threat that recently urbanized areas of the county faced, Reagan persuaded the Board of Supervisors to put a twenty-seven-million-dollar bond measure on the November 1926 ballot to build protective works for these newly vulnerable areas. The Municipal League, for its part, was determined to focus the election on Reagan. Not only were the bonds on the ballot, but Prescott Cogswell, the supervisor from San Gabriel Valley, was up for reelection. He was a staunch supporter of the Forks site and was one of the three-member majority on the board that supported Reagan. Since Reagan served at the pleasure of the supervisors, the Municipal League saw the elections as an opportunity not only to defeat the bonds but also to oust Cogswell and therefore perhaps Reagan.

Along with the *Los Angeles Record* and the *Hollywood Citizen,* the Municipal League's *Bulletin* attacked Reagan. He "had no training as engineer or geologist," his critics charged, and they mocked the folly of entrusting construction of the world's tallest dam to someone who had never built one before. They pointed out that *Who Is Who in Engineering* had not listed him among the eighteen thousand engineers, including four hundred in Los Angeles, whom that publication cataloged. And they accused him of staffing the Flood Control District with incompetent cronies as payment for personal favors and friendships.[47] To defeat the bonds and discredit Cogswell, for whom Reagan was campaigning, the Municipal League portrayed flood control under Reagan as an inefficient operation headed by corrupt incompetents who made decisions without public oversight and wasted public money on expensive projects to which cheaper alternatives existed. In the eyes of the league and its allies, flood control seemed to be a nightmare of policy-making disorder.

The voters of Los Angeles County and the federal San Gabriel River Commission both passed judgment in November 1926, but they rendered opposite verdicts. The federal government awarded a right-of-way for the Flood Control District's railroad, tacitly endorsing the San Gabriel Dam project and the Forks site in particular. In his report to the secretary of the interior, the head of the commission, Frank Safley, blasted the "insidious propaganda" campaigns to discredit Reagan and block the project.[48] The electorate, however, accepted the propaganda. In this election, without a drought to magnify the extremes of deluge and scarcity, it was Reagan and his project, not the rivers, that appeared to be in need of reengineering. Not only did voters reject the bonds, but they also sent Cogswell packing, replacing him with a candidate who had promised to reorganize the Flood Control District.[49]

The effects of the campaign rippled beyond the election. The most immediate consequence was the reorganization of the flood-control leadership. The newly constituted Board of Supervisors took immediate steps to allay the criticisms of the district. Hoping to determine once and for all where and if San Gabriel Dam should be built, the board sought new experts. They replaced Reagan with E. C. Eaton, who had distinguished himself working for the California state engineer's office. They also impaneled a consulting board of engineers to assist Eaton, naming Reagan as one of its three members. At the same time, the Municipal League, the California Taxpayers Association, the American Society of Civil Engineers, and the Los Angeles Chamber of Commerce (which now opposed the Forks site after having supported it in 1924) pressured the supervisors into appointing an independent team of engineers from outside the region to compare the Forks and Dike sites.

The team's report the following March pleased no one. While rejecting the Granite Dike, the consultants also criticized the Forks plan, saying it would cost considerably more than the Flood Control District had budgeted. They suggested instead that the district build a smaller dam there. The bewildered supervisors abandoned the high-dam plan and prepared instead to go ahead with a smaller, 385-foot structure at the Forks.

San Gabriel Valley residents reacted angrily. They filed suit for an injunction to block the new lower dam and demanded that the supervisors reauthorize the 425-foot structure. The California Taxpayers Association filed a counter suit to block the high dam. In March 1928, Judge Anderson of the county superior court settled the matter exactly as the Flood Control District's lawyers had predicted back in 1925 when they warned the supervisors against the Dike site. He accepted the high-dam advocates' claim that the Flood Control Act's "change of conditions" clause required a change of "physical" conditions before the Flood Control District could abandon the plans approved by voters in 1924. So far, he said, none of the competing reports indicated anything to constitute such a change. He ordered the district to proceed with the high dam at the Forks site.[50]

In July 1928, the supervisors finally began a search for bids for the construction of San Gabriel Dam.[51] By the time they selected a contractor in November, the machinery of public policy had been thoroughly gummed up. The experts had failed to agree; the public divided against itself; and the law provided no way to resolve the disputes until Judge Anderson decreed that no conditions had changed in the canyon. After five years, one federal hearing, two elections, two lawsuits, and numerous conflicting engineering reports, the Flood Control District had yet to move its first shovel of dirt in San Gabriel Canyon. Politics made engineering a messy process.

FAULTY ROCKS, FAULTY POLITICS

Right at the moment that the high-dam/low-dam controversy finally seemed to be resolved, the rupture of a different dam set in motion events that would soon affect the fate of the project in San Gabriel Canyon. Just after midnight on 12–13 March 1928, St. Francis Dam in northwestern Los Angeles County began to crack. Shortly after filling to capacity for the first time, the dam that many had hailed as a marvel of engineering suddenly burst. As the side sections crumbled, water crashed down the Santa Clara River, and concrete chunks weighing several thousand tons each bounced for half a mile down the channel. The torrents smashed a powerhouse and destroyed a construction camp where more than a hundred workers slept. In the early hours of the morning, the flood reached an agricultural valley, where it destroyed more than twelve hundred homes and numerous farms and town buildings. By six o'clock, when the waters returned to their bed, at least four hundred people had lost their lives. Crumpled masses of concrete thirty feet by fifty feet protruded from the meandering stream, and only a hundred-foot-wide center section of the dam stood on its original foundations, a lonely testament to the structure that had towered there.[52]

St. Francis Dam had nothing to do with flood control, yet it had significant impact on that program. St. Francis Dam stored aqueduct water from the Owens Valley for the Los Angeles Bureau of Water Works.[53] Ironically, a structure designed to prevent water shortage caused the deadliest flood in southern California history. No earthquake or torrential storm had triggered the failure, only the construction of the dam on unsolid foundations. St. Francis Dam stood less than half as high as the proposed San Gabriel Dam and stored only one-sixth the volume of water. Occurring three days after Judge Anderson directed the Flood Control District to build the high Forks dam, the St. Francis catastrophe spurred engineers to take a closer look at San Gabriel Dam's foundations, an investigation that soon rendered the five years of haggling that climaxed in Anderson's courtroom largely irrelevant. The struggles between 1924 and 1929 had not killed the project but rather merely delayed it. There was, however, one more twist in store for the star-crossed project. Between 1928 and 1934, unrelated events, starting with the St. Francis disaster, revealed physical flaws and political corruption that would eventually doom not only the San Gabriel project but the entire local flood-control regime.

Geology uncovered the structural flaws. The same geological processes that raised the mountains and filled the valleys with water-storing gravels and fertile topsoil also left little bedrock stable enough to anchor a dam the size of the San Gabriel. The uplift of the mountains began with movement along faults in the late Pliocene and early Pleistocene eras and continued slowly for several hundred thousand years. The forces of uplift, however, were uneven, causing the rock to shear in weak places as the mountains rose. The rock in these zones fractured into irregular blocks, anywhere from a few inches to several feet in diameter. In some places it had been crushed into small pieces, and in others it had disintegrated entirely. Rivers seeking the path of least resistance followed along the easily eroded sheared zones. The east and west branches of the San Gabriel River, which meet at the Forks, overlie the San Gabriel fault, which divides the northern and southern blocks of the range. The main branch of the river, which runs south from the Forks, also follows a fault. Thus the Forks, the proposed site of the world's greatest dam, lay right on top of two earthquake faults.[54] To build a dam in San Gabriel Canyon necessarily meant putting it on a weak foundation.

Few people in the 1920s, however, knew this. Engineering geology was an infant science, and few people had even looked at the geology of the Forks site.[55] When Reagan had first investigated San Gabriel Canyon, an old-timer who had built an aqueduct in the canyon cautioned him against the Granite Dike site. Rock at the Dike site was fractured by faults; some fissures were big enough to drive a streetcar through them. "Stay away from that country," the man warned; "it is all slipping."[56] Although Reagan rejected the Dike site for these reasons, he believed the Forks location to be solid. As he testified to Anderson's court, he had learned much about canyon geology in his career, and he had walked over every square foot of the Forks site without finding any evidence of faults. He saw little reason to investigate further. Nor did he ask others to do so. The Flood Control District consulted Davis, Goethals, and others on financial and engineering questions only. Not even Reagan's opponents sponsored geologic studies of the sites.[57] Not until the district commissioned the University of California geologist Andrew Lawson in 1925 did anyone do a geological study of the site. Even Lawson spent only two days and charged only $355 before pronouncing the site stable. Faults underlay the site, he said, but they were not active. Even if they were, he declared, the movement could not break the dam. The surrounding mountains "would have to be destroyed before the dam could be demolished."[58] In all the studies and counterstudies that the San Gabriel Dam politicking had produced, engineers had considered the dam height, location, cost, and reservoir capacity of the site, but no one had taken a good look at the rock underneath it. Even in the aftermath of the St. Francis disaster, the $25 million San Gabriel Dam had only a $355 study attesting to its geological safety.

It appears that only the contractors hired in 1928 knew of the geological hazards of the site. In 1927, in preparing their bid on the San Gabriel project, they had retained a civil engineer, Allan Sedgwick, who reported that the faults and shear zones at the Forks cast doubt on whether the dam would stand or could even be built at all. "Evidently," he indicated, "a thorough geologic study of these conditions has not been made and certainly such a study should be made before a dam of the tremendous size of the San Gabriel project is built." Failure of such a dam, he warned, "might be very disastrous."[59] The contractors later maintained they had given a copy of this report to the Board of Supervisors, but the supervisors' copy that surfaced a year later in 1929 was missing the four crucial pages that described Sedgwick's concerns. In the meantime, no geologic report was undertaken, and the contractors proceeded with excavation of the site.

Other little-noted irregularities also attended the construction. The contractors had what was known as an unbalanced contract, meaning they had organized their bid in such a way as to earn the bulk of their profit early in the construction process. They would do the excavation work at considerably more expensive rates than other contractors offered, but then they would built the dam at cost. With both a damning geology report and a front-loaded contract in their possession, they dynamited the canyon sides through the summer of 1929 and removed the rubble, earning $2.95 for each cubic yard of material they excavated—almost five times the standard cost of such work. The county was now overpaying excavators to construct an unbuildable dam on an unsuitable foundation under a court order and a flood control act that required the project to go forward unless there was to be a "change of conditions."

On 16 September 1929, conditions changed. The west abutment at the dam site caved in, and five hundred thousand cubic yards of mountainside slumped into the canyon. Only a month before, in response to the St. Francis failure, the California legislature had enacted a statute requiring state approval of all dams. Although the state had previously taken no regulatory role in Los Angeles flood control, the new law required state inspection and authorization of plans for all dams to be built within California, including those of the Flood Control District. The Board of Supervisors had not yet submitted the San Gabriel Dam plans for approval. Acting under the new code, the state engineer Edward Hyatt appointed six engineers to inspect the geology in the vicinity of the Forks. The panel declared the site unsafe. On 26 November, Hyatt informed the Board of Supervisors that San Gabriel Dam would be "a serious menace to life and property," and he prohibited further construction.[60] The disaster triggered by Los Angeles's efforts to secure a water supply rippled outward to alter, in concert with the geology of the Forks site, the fate of San Gabriel Dam. Nature and city building combined to thwart southern Californians' attempts to control the flow of water.

The slide, however, revealed more than just faulty ground; it also bared a chain of corruption that discredited the flood-control program as a whole. At first the scandal unfolded without fanfare. In October 1929, the Sedgwick report, this time with the critical pages intact, surfaced and was compared against the supervisors' mutilated copy. In November, an insurance company that held a policy on the dam refused to return the outstanding money on the policy to the board. In January 1930, the contractors threatened to sue the district for lost profit. Anxious to close the matter and perhaps a little leery that they might be found negligent for not having had the state inspect the dam earlier, as the St. Francis Dam law required, the supervisors ignored the incriminating Sedgwick report, allowed the insurance company to keep $30,000 to which it was not entitled, and settled with the contractors for a whopping $830,000.[61]

The muckraking *Daily News*, however, smelled a scandal. As reporters began investigating both the thirty thousand dollars and the mysterious Sedgwick report, they traced the money through a series of shady deals to Supervisor Sidney Graves's former assistant, "Chicago Harry"

Merrick. The ensuing grand jury investigation also uncovered the front-loaded contract, as well as an eighty-thousand-dollar bribe the contractors had paid to Graves to secure the January 1930 settlement. In an October 1930 report, the grand jury's investigators upbraided the Board of Supervisors for careless use of public funds and recommended further investigation. Eventually, the grand jury indicted Graves and the new Flood Control District chief Eaton and accused the contractors of fraud. At no time, the grand jury found, did the contractors intend to complete the dam, but rather had accepted the front-loaded contract knowing that the dam could not be built. The grand jury directed the Board of Supervisors to bring suit to recover not only the $830,000 settlement but also the $1.8 million it had previously paid for excavation. Eaton was later acquitted. Graves eventually went to the state penitentiary at San Quentin. The matter with the contractors was never fully resolved, however, as the board eventually accepted $738,000 from the contractors in an out-of-court settlement in 1936, thus drawing to a close the San Gabriel Dam fiasco.[62]

Born of an unlikely combination of circumstances in which drought and urbanization momentarily realigned flood-control politics, San Gabriel Dam died an equally sudden and unexpected death—ironically, at a point when, for the first time, no political obstacles blocked its path. Federal right-of-way conflicts, engineering disputes, lawsuits, and countersuits had been cleared away. All signs in 1928 had pointed to the dam's prompt completion. The St. Francis Dam failure and the San Gabriel construction-site landslide, however, changed all the rules. By provoking a state investigation, by inducing the county to reach an ad hoc settlement with the contractors, by raising questions about why no one had investigated the geology, and by requiring the Flood Control District to account for its spending, the two unrelated events revealed all the previously invisible irregularities surrounding the San Gabriel Canyon project. As a result, through the 1920s, the flow of water in the canyon remained, despite the best efforts of engineers, beyond human control.

A TWISTED PROGRAM

In 1929 the Municipal League demanded to know, "What twisted a well-thought-out damless program of $16,000,000 into one of $67,000,000 largely made up of concrete dams?"[63] As usual, the league slanted the truth a bit. The Board of Engineers' sixteen-million-dollar comprehensive plan of 1915 had not been literally "damless," but rather had recommended fewer and smaller dams than the program Reagan later commenced in 1924. Also, Flood Control District spending never reached sixty-seven million dollars in the 1920s; that was merely the total amount of bonds Reagan had requested. The Municipal League was nevertheless correct in its assessment that by the end of the decade, the flood-control program had traveled a tortuous path. Departing from the sixteen-million-dollar program the Board of Engineers had outlined in 1915, the Flood Control District had undertaken a thirty-five-million-dollar quest to build a set of big dams, including the world's tallest. It had greatly enhanced the role of water conservation in flood-control plans. And five years of political struggle had yielded no increase in flood protection. In 1929, the Municipal League pronounced the flood-control plan a boondoggle and blamed both Reagan's ambitions to build great engineering monuments to himself and the "greedy, overreaching big business" concerns such as the cement and steel companies that stood to benefit from the big dam.[64] Although both corruption and Reagan's ego did influence the district in the 1920s, neither was the basis for the decision to embark on the quixotic project nor the chief source of its failure.[65]

Ultimately the district's setbacks in the 1920s stemmed from ordinary features of the urban ecosystem. Southern Californians were trying to replace ancient hydrologic processes with a human technological order. That new order encompassed not just wells, dams, ditches, and plows but also an entire complex of economic, political, and physical relationships that injected new sources of disorder into the flood-control efforts. The 1924 election coincided with drought and economic growth to inspire demand for a previously suspect project. The same sort of appeals to expertise and public interest that in the 1910s had successfully launched the flood-control program later produced stalemate once experts and citizens began to disagree in the 1920s. Thus the timing of unrelated events and changing environmental and political contexts exerted as much influence on how water would flow as engineering did. At the end of the 1920s, water still flowed freely in San Gabriel Canyon, not because of the awesome force of nature, but rather because ordinary features of city building—such as municipal elections, water conservation, and the language of laws—combined with the region's normal geologic and climatic characteristics to thwart flood-control plans.[66] The thesis of hydraulic order had yielded the antithesis of flood-control policy-making chaos.

At first the Flood Control District's troubles had little impact on flood prevention. Another long dry phase had followed the 1916 flood, and no major flood struck during the 1920s. During that period, however, the construction of flood-control works failed even to keep pace with urban development, so that by the 1930s, many of the same conditions that had produced the 1914 flood still prevailed, and many recently developed places lay unprotected. An overlooked by-product of the San Gabriel Dam fiasco was the 1926 bond election, in which fed-up voters defeated a proposal for flood-control works in the recently suburbanized La Cañada Valley, a smaller watershed to the west of San Gabriel Canyon. Through the 1920s, then, urbanization waxed while the technocratic juggernaut waned. It was a formula for disaster.

THE EARLY RIPARIAN LANDSCAPE

Figure 1. Rubio Wash, northeast of Los Angeles, 1914. In the late nineteenth and early twentieth centuries, riverbeds were ill-defined. Streams spilled from San Gabriel Mountain canyons into gravelly washes that were filled with vegetation and constantly shifting channels. (Photo courtesy of Water Resources Center Archives, University of California, Berkeley.)

THE EARLY DEVELOPED LANDSCAPE

Figure 2. La Cañada Valley, after 1887. By the late nineteenth century, farms and homes were scattered across the fertile debris cones at the foot of the San Gabriel Mountains, with canyons pointed right at them. Compare with figure 8. (Photo courtesy of California Historical Society/TICOR Collection, Specialized Libraries and Archival Collections, University of Southern California.)

Figure 3. Pacific Electric Railway trestle across Los Angeles River, between downtown and Long Beach, early twentieth century. In the late nineteenth century, development crept into the apparently harmless washes on the coastal plain. Farmers plowed land and built homes and other structures next to and sometimes in the washes. Railroad trestles crossed the rivers, blocking the channels. By the early twentieth century, a new urban landscape was emerging, superimposed on the old riparian landscape—though the latter was hardly obliterated, as southern Californians would discover in 1914. (Photo reproduced by permission of The Huntington Library, San Marino, California.)

THE 1914 FLOOD

Figure 4. February 1914 flood, Los Angeles River, eight miles southeast of Los Angeles. In the 1914 flood, the rivers smashed bridges, inundated towns and farms, filled the harbor with silt, and swallowed railroad cars, unmasking all the previously invisible hazards of building a city on the floodplain. (Photo reproduced by permission of The Huntington Library, San Marino, California.)

Figure 5. Reinforced bank of Los Angeles River, 1938. To stabilize the banks, channels were lined with rocks and wire, as shown here. Elsewhere, wooden pilings and even old automobiles were used. (Photo courtesy of California Historical Society/TICOR Collection, Specialized Libraries and Archival Collections, University of Southern California.)

Figure 6. Site of proposed San Gabriel Dam, looking downstream, 1999. In the mountains, the Flood Control District planned to build large dams in the 1920s, some of which were never built because political conflict made engineering just as unstable and unpredictable as the landscape the district was trying to rationalize. The dam at this site would have created an enormous reservoir, big enough, engineers hoped, to store water for dry seasons and to catch floods in wet ones. (Photo by the author.)

Figure 7. Check dams in San Gabriel Mountains, 1910s. In smaller canyons, the Flood Control District built hundreds of rock and wire structures, eight to ten feet high, turning some canyons into hydraulic staircases that slowed water and captured debris. (Photo courtesy of Water Resources Center Archives, University of California, Berkeley.)

Figure 8. La Cañada Valley, 27 January 1934. By the 1930s, the foothills of the San Gabriels were a very dangerous place. Canyons, with years of boulders, mud, and other debris stored in them, were aimed right at the suburban communities that had mushroomed downslope on the debris cones during the preceding two decades. On New Year's Eve 1933–1934, the canyons poured water and mud over Montrose and other foothill towns. (Photo reprinted by permission of County of Los Angeles Department of Public Works.)

Figure 9. Flood damage to neighborhood, Montrose, La Cañada Valley, 27 January 1934. On the morning of 1 January 1934, residents of Montrose and other communities awoke to find their yards and streets covered with mud. (Photo reprinted by permission of County of Los Angeles Department of Public Works.)

Figure 10. Flood damage to American Legion Hall, Montrose, La Cañada Valley, January 1934. The waves of boulder-laden liquefied mud smashed power lines, cars, and homes. Twelve people died in the American Legion Hall, the building pictured here with the curved roofline and stone chimney. (Photo courtesy of Department of Special Collections, Charles E. Young Research Library, UCLA.)

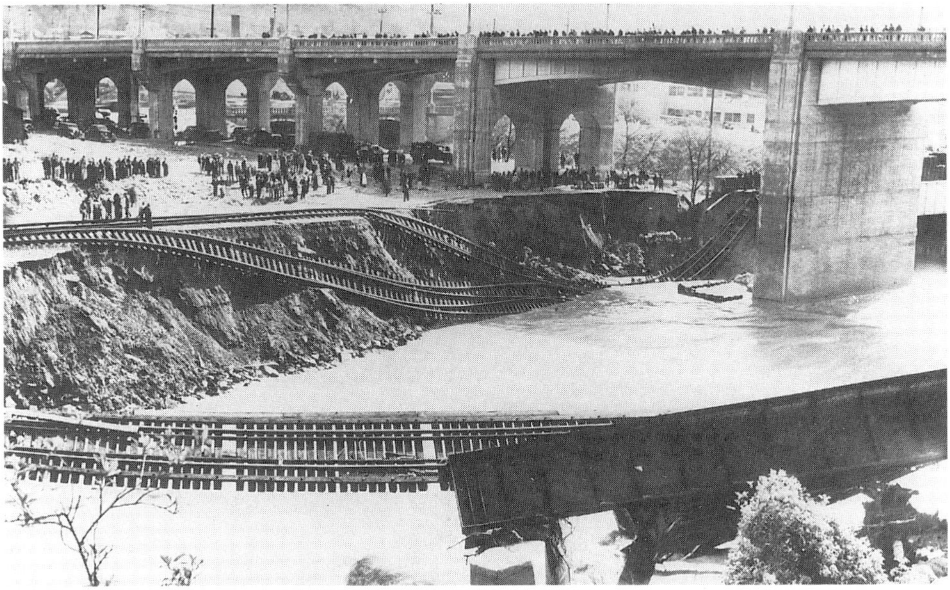

Figure 11. Flood of 1938, Los Angeles River, north of downtown. The 1938 flood was the worst in recorded southern California history. Here, the Los Angeles River has twisted the steel tracks of a Southern Pacific Railroad trestle near Elysian Park. (Photo courtesy of Los Angeles Examiner Collection, Specialized Libraries and Archival Collections, University of Southern California.)

Figure 12. Flood of 1938, Anaheim. The rivers flooded towns in Los Angeles and Orange counties. Here, southern Californians take to their beloved automobiles to flee the waters of the flood. (Photo courtesy of Los Angeles Examiner Collection, Specialized Libraries and Archival Collections, University of Southern California.)

Figure 13. Flood of 1938, Venice Beach. Not wanting to miss an opportunity to catch a good wave, a pair of beach boys hitched their surfboards to an automobile during the flood. This new sport did not catch on. (Photo courtesy of Los Angeles Examiner Collection, Specialized Libraries and Archival Collections, University of Southern California.)

Figure 14. Model Yard of the U.S. War Department, 1940. In contrast to Venice's youth, few southern Californians saw the 1938 floods as a recreational opportunity. To protect the city, the engineers went back to their drawing boards, envisioning bigger, stronger devices to control the floods. To test their designs, the engineers constructed scale models. Here the model Arroyo Seco flows into the model Los Angeles River, at a point that would be just north of downtown. (Photo courtesy of Water Resources Center Archives, University of California, Berkeley.)

Figure 15. Paving the bed of the Los Angeles River south of downtown, 21 September 1951. In the beds of the Los Angeles River and other waterways, models became reality through the use of cement, steel, and machinery. (Photo courtesy of the National Archives, Pacific Region, Laguna Niguel, California.)

Figure 16. Los Angeles River, 12 September 1940. When all the reconstruction was complete, the formerly dusty, vegetation-choked beds were fields of concrete. The low-flow channel, at the center, kept a ribbon of water trickling fast enough to prevent sediment deposits from choking the reconstructed bed, in the way they had frequently filled the prepavement channels. (Photo courtesy of Water Resources Center Archives, University of California, Berkeley.)

Figure 17. Whittier Narrows Dam and Flood Control Basin, 30 June 1955. Whittier Narrows and five other flood-control basins barricaded the rivers. Sandy and dry, and apparently pointless most of the time, their reservoirs filled during floods and funneled water downstream at a rate that the channels could in theory handle. (Photo courtesy of the National Archives, Pacific Region, Laguna Niguel, California.)

THE 1980 FLOOD

Figure 18. Flood of 1980, Los Angeles River, San Fernando Valley. This box channel is fifty feet wide and sixteen and a half feet deep, and yet the Los Angeles River managed to fill it. LACDA survived the storm—barely. (Photo reprinted by permission of County of Los Angeles Department of Public Works.)

Figure 19. Flood of 1980, Los Angeles River levee, near Wardlow Road. This is the infamous debris that the Los Angeles River deposited on its levee top, leading the Army Corps to reevaluate LACDA's adequacy. (Photo reprinted by permission of County of Los Angeles Department of Public Works.)

THE REEVALUATED LANDSCAPE

Figure 20. Los Angeles River, San Fernando Valley, near Universal Studios, 2000. The dimensions of the box channel at this spot are similar to those where the channel filled in 1980 (see figure 18). Behind all the fences and barbed wire, the clean geometry of the riverbeds and low-flow channels created an aesthetic of artifice hidden from most southern Californians. (Photo by the author.)

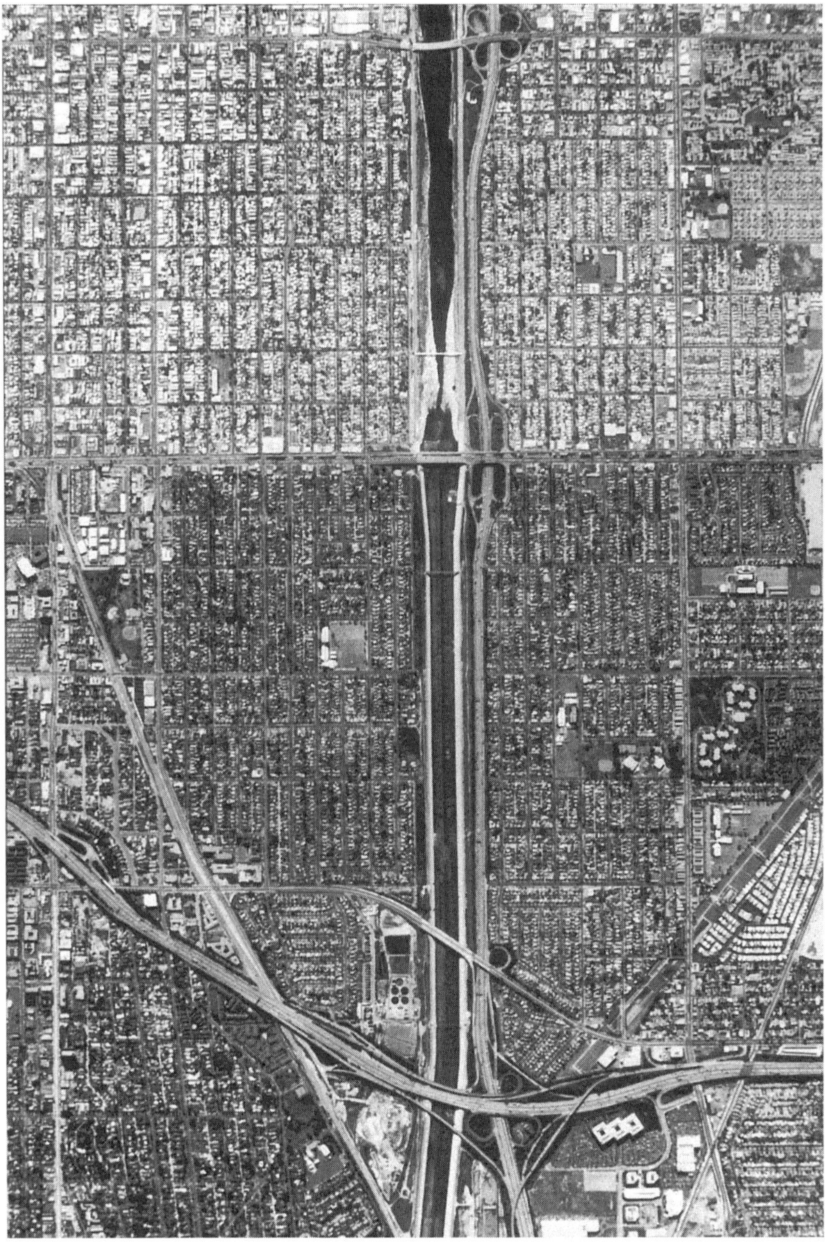

Figure 21. Aerial view of Long Beach, late twentieth century. Here, at the Long Beach Freeway–San Diego Freeway–Los Angeles River interchange, twentieth-century Los Angeles flood control reached its ultimate expression. The channel was straightened, paved, and confined, and development left hardly a square foot of landscape untouched. And yet as late as 1992, this spot was still imperiled by a hypothetical flood that engineers feared might overtop the banks and do as much as two billion dollars of damage to the urbanized landscape. (Photo reprinted by permission of County of Los Angeles Department of Public Works.)

Figure 22. Ernie's Walk, Los Angeles River, San Fernando Valley, 2000. Ernie's Walk, the first of what came to be dozens of pocket parks, beautified stretches of the river and illustrated future possibilities for using the waterways to revitalize the city. (Photo by the author.)

Figure 23. Arcadia Wash, April 2000. Amid this massive hydraulic transformation, ubiquitous small channels like this one in a suburb fifteen miles east of Los Angeles blend right into the early twenty-first-century southern California landscape. Although they look and act quite differently from the channels of 150 years before, the rivers remain integral features of the urban ecosystem. (Photo by Richard J. Orsi.)

CHAPTER 4

A More Effective Scouring Agent

The New Year's Eve Debris Flood and the Collapse of Local Flood Control, 1930–1934

In late December 1933, a wet warm front advanced from its tropical origins northeastward to the California coast. Cyclonic weather patterns in December are not uncommon in the area, and at first nothing distinguished this one as it showered coastal southern California with moderate rains beginning on the afternoon of the thirtieth. But cold winds blew from the east, undercutting the tropical storm. The wet warm air mass rose before it reached the mountains, and it could no longer hold its moisture. Rains fell over the foothills in intensities more frequently experienced at the ridge crests. In eleven hours, the storm cell delivered eleven inches of rain at elevations between two thousand and five thousand feet above sea level.[1]

The rains had arrived belatedly that year. As late as November, wildfires crackled through the desiccated chaparral of the southern foothills. One blaze that began in Pickens Canyon above the La Cañada Valley on 21 November denuded a 7.5-square-mile swath of watershed.[2] When the rains finally came in a two-day storm on 14–15 December, they quickly washed the fire's detritus off the hillsides and into the canyons. As rain again pelted the foothills at the end of the month, water cascaded over the ground, gathering soil and rubble. In the canyons, debris from the fire and material impounded by check dams soaked it all up. On the evening of 31 December, things began to slip.[3]

The rain did not dampen the plans of most New Year's Eve revelers that night. In the foothill community of Montrose, party-goers filled the

American Legion Hall for a year-end soirée, while others left their children and gathered at the homes of neighbors. There was little celebrating, however, at the home of Harold Hedger, the office engineer for the Los Angeles County Flood Control District. Hedger and his family were among the throngs of people who had moved to the La Cañada Valley during the previous decade. With the careful eye of an engineer, he had selected a home site on high ground—safe from floods, he reasoned—and to that haven his neighbors flocked that night, seeking shelter from the rain and the rising waters. Late in the evening, he was called to the Flood Control District headquarters, where incoming reports were indicating that streams were swelling all over the county. Hedger kept in touch with his family and neighbors by phone for several hours while directing the district's emergency efforts. Meanwhile, the reports from the La Cañada Valley were growing more worrisome. At one point, Hedger was speaking on the phone with someone in the area, and he heard a scream on the other end. The connection went dead. Hedger tried frantically to reach the county sheriff in Glendale, just down the hill from the stricken valley, but he was unreachable that New Year's Eve, as was the Glendale police chief. Just before midnight, Hedger dialed his own number. That line was dead too. For the next twelve hours, he would have no idea whether his family was safe or not. When he finally reached someone at the sheriff's office, he could not convince the person that a strange disaster of flowing mud was unfolding just up the hill. No one was expecting what Hedger was describing—it had never happened before.[4]

But something was happening now. As the rain fell even harder around midnight, roars echoed from the canyon mouths. Mountain slopes slumped into the gorges. Torrents gathered the debris into viscous globs that oozed down the canyons. They picked up boulders and tree branches, churning them into a consistency that resembled wet concrete. The numerous check dams offered little resistance to the flows, which snapped the puny structures or simply rolled over them, collecting the material accumulated behind the barriers and then continuing down toward the valley. As the masses rumbled down the canyons, tributaries delivered more rubble, sometimes momentarily clogging the main channels until enough water gathered behind the bottleneck to saturate the masses and restore their advance. In this way, the masses gathered more and more debris and water until waves of muck fifteen feet and higher crashed out of the canyons, overflowed the creek channels, and spread hundreds of feet wide.[5]

The debris stormed first out of Blanchard Canyon, at the northwestern end of the La Cañada Valley, shortly before midnight. At 12:09 A.M., Shields Canyon erupted with a wave of mud that a resident watched as it crossed Foothill Boulevard. Three minutes later it smothered his home, which lay six hundred feet from the channel. Nearby, the billows pinned a man against an automobile and then lifted both across the street. Mud rushed over his head and buried him up to his neck. To the southwest, debris from Pickens Canyon smashed through the upstream wall of the American Legion Hall, killing twelve of the revelers before crashing out the downstream side of the building. Hearing a rumble, one party host went to his front door, just in time to see his front porch wash away. "Everybody out!" he warned his guests—but his cry was too late to save five of them. Blocks away, a twelve-year-old girl baby-sitting her two younger brothers awoke to the sound of logs banging against the side of the house. As the sludge poured in, she roused the two boys, and the three clambered up into their backyard tree house minutes before the whole home collapsed. All over the valley that New Year's Eve, the deluges rampaged. Power lines toppled. Bridges collapsed. Automobiles piled in wrecked tangles. Mud covered everything. And then it was gone. By half-past midnight, residents near the canyon mouths could walk across the streams from which such terror moments earlier had issued.[6]

The Hedgers' home escaped unscathed, but more than four hundred other homeowners in the valley were not so lucky. Even less fortunate were the forty people who lost their lives that night. In the end, the deluges spread debris over a quarter of the valley's seven-square-mile area. In less than one hour, the debris flows moved more than six hundred thousand cubic yards of earth out of the mountains and into the valley.[7] "It is hard to imagine," a pair of foresters observed, "a more effective scouring agent than this muddy boulder-laden flood."[8] Indeed, the flood had moved at least a hundred thousand cubic yards more than the Los Angeles County Flood Control District had managed to excavate in five years at the San Gabriel Dam site.

The 1914 vision of hydraulic order had mutated into a technocratic nightmare. Over the previous twenty years, southern Californians had spent fifty million dollars on a comprehensive plan, a centralized authority, and a giant concrete weir, all in the hopes of reengineering the local hydrology. Despite the expensive and ambitious program, a new type of flood disaster, this time involving mud in the foothills instead of water on the plain, struck on New Year's Eve 1933–1934. When southern

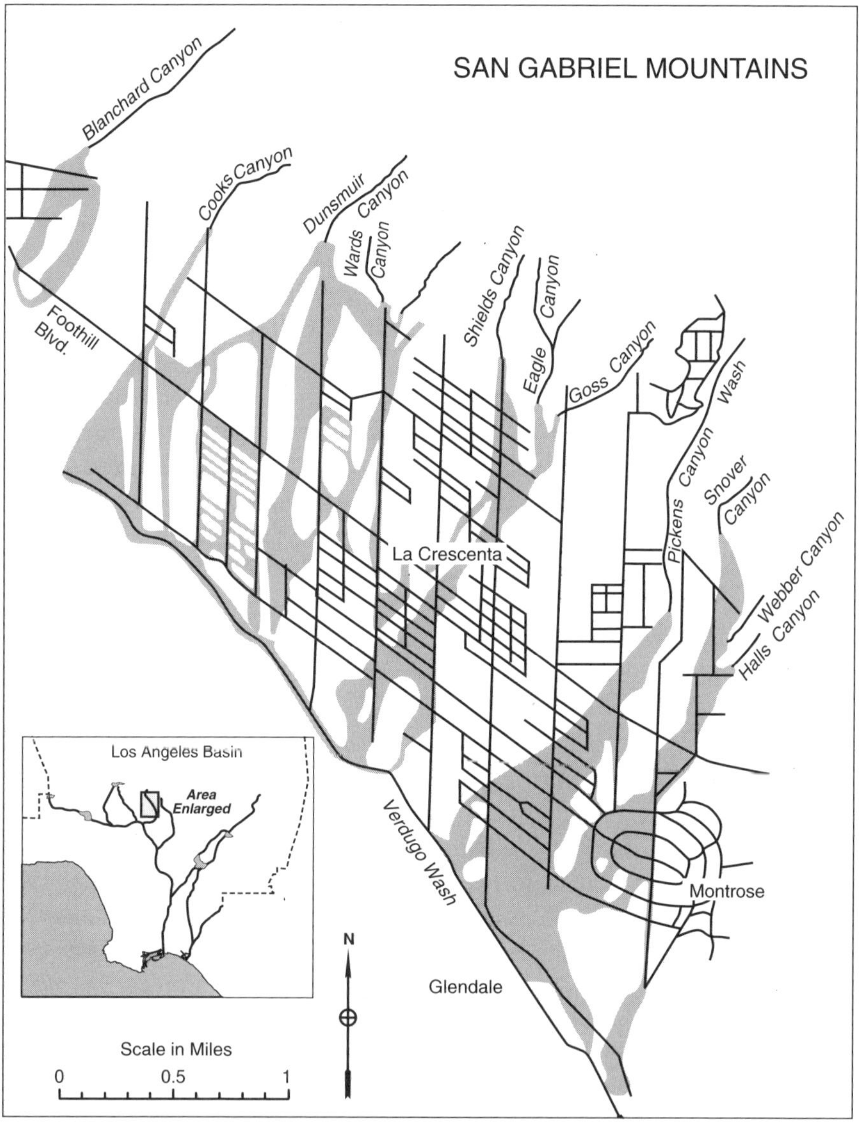

Map 5. La Cañada Valley, 1934. The shaded portions indicate the suburban areas inundated during the New Year's Eve debris flow. See also figures 8–10 in this book's photo section. (Map by Carol Marander.)

Californians sought to assign blame for the tragedy, they pointed at the extraordinary combination of fire and rain.

The La Cañada disaster, however, was a failure of the entire urban ecosystem, not merely the weather. Just as they had since 1915, technocrats of the late 1920s and early 1930s concocted flood-control policy from a mixture of science, politics, and economics. At the same time, idyllic foothill subdivisions lured people up the slopes to settle in harm's way. Even after the New Year's Eve debris flow, the local political and economic conditions of 1934 continued to prevent engineers from redirecting the district's energies and resources to respond to this latest challenge, and the local flood-control regime toppled. Individually, the rain, fire, mud, flood-control plans, suburban development, and political and economic conditions were unremarkable features of the urban landscape of Los Angeles in the 1930s. Together, however, they constituted a system within which something went terribly wrong on New Year's Eve. It was the interaction of the ordinary, not the violence of the extraordinary, that produced the new type of disaster and brought an end to the first phase of flood control in southern California. Technocracy, city growth, and nature joined to make Los Angeles a hazardous metropolis.

A NEW COMPREHENSIVE PLAN

In 1934, water ran through the American Legion Hall, but that is not where E. C. Eaton had planned it to flow. Eugene Courtlandt Eaton, or "Court," as he was called, had replaced James Reagan as Flood Control chief in early 1927, amidst great controversy, and provided a stark contrast to his flamboyant predecessor. His impeccable civil engineering credentials included a degree from McGill University, in Montreal, and twenty years of experience in construction, including six years as the engineer in charge of dam building for the California state engineer's office in the 1920s. He had written acclaimed articles on various civil engineering topics, and he would publish more before his career was over.[9] Harold Hedger, who worked under Eaton for seven years, later characterized him as a sincere man and a good engineer but "politically inept."[10] Whereas Reagan had dreamed of great mountain dams large enough to catch any flood imaginable, whatever the cost, Eaton was a realist. He programmatically believed that a "complete ultimate solution must be undertaken progressively, limiting expenditures to locations as property values warrant, and conservation where and when benefits will accrue as a whole."[11] Reagan had been daring and impulsive. Eaton was cautious and methodical. Shortly after taking over the Flood Control District, Eaton shifted its program to a strategy more befitting his meticulous engineering style.

Eaton inherited from Reagan a Flood Control District in disarray. Even Reagan's protégé and admirer Harold Hedger remembered his mentor's flood-control program as "piecemeal." Reagan, Hedger said, would decide he wanted a dam in one canyon or another and then design it with only cursory examination of hydrologic and geologic conditions. By the time Eaton took over, the program was little more than a sequence of poorly done and often temporary channel work and a collection of dams, each of which attempted to meet flood-control needs in a given corner of the county but which had little relation to one another or to any general countywide strategy.[12] Making matters worse, voters had just defeated the 1926 bond issue to finance additional construction, including badly needed works in La Cañada Valley. Reagan resigned after a swing in the balance of power on the Board of Supervisors, a February 1927 flood that washed out the Pacoima Dam construction site, and a dispute with one of his top assistants about district management.

Meanwhile, as Los Angeles boomed in the 1920s, the flood problem "multiplied appallingly," in the words of Flood Control District officials.[13] In the decade and a half since the passage of the 1915 Flood Control Act, the population in the county had nearly tripled, and the assessed value of property had quadrupled. This development continued the processes that had contributed to the flood hazard prior to 1914. As automobiles and paved streets replaced wagons and muddy thoroughfares, impervious surfaces increasingly covered the basin. By 1931, the surface area of pavement totaled more than fifty square miles. Runoff, which had averaged 20 percent of precipitation when it fell largely on unpaved surfaces, now surpassed 80 percent. Neighborhood subdivisions added to this problem, as urban development covered formerly agricultural areas and other open spaces. In the foothills, residential developments now covered the same debris cones over which water had previously meandered in various channels. On the plains, channels had to carry not only their own flows but also the runoff from broad swaths of land whose waters the storm sewers concentrated and delivered to the stream channels. With urban development, less water ponded and puddled and percolated into the ground. More water from a larger area flowed faster and took less time to concentrate. Channel improvements in one place merely increased the possibility of a break elsewhere. A dam in one canyon meant less money for flood control in some other locale. As it became evident that the metropolis was going to envelop the entire basin from the foothills to the coast, the piecemeal approach grew more inadequate. Soon, flood control would be needed everywhere.[14]

To this end, Eaton undertook the formulation of a new plan that would coordinate all projects in the district and prioritize them according to their urgency and economic feasibility. For four years, Eaton's engineers and surveyors studied old hydrologic records and gathered new ones, taking measurements at 310 rainfall stations, 150 stream gauges, 2,700 wells, 57 evaporation and snow stations, and 4 debris stations.[15] From this data, they drew up a picture of the Los Angeles Basin hydrology and used it to coordinate a plan for controlling of the flow of water all over the district.

In September 1931, Eaton unveiled the report, which came to be known as the Comprehensive Plan and which outlined all of the works that would eventually be needed on all of the stream channels and in all of the canyons. Shifting away from the 1920s emphasis on great dams, the Comprehensive Plan proposed numerous low dams in the mountain canyons to slow the floodwaters. The waters would be conveyed to gravel pits and other porous spreading grounds to percolate into the enormous groundwater basins that underlay many parts of the county. The San Gabriel project, aborted once the state supreme court ruled in 1931 that the 1929 landslide constituted a "change of conditions," was split into a pair of smaller dams, with spreading grounds to make up for the lost storage capacity.[16] Low dams with reservoirs behind them would be built on the valley floor. Later known as flood-control basins, these structures would remain empty except during flooding, when they would catch upstream floodwaters and release the water downstream in a controlled way. Any excess water after the flood would be conveyed to the spreading grounds. Channels between the dams would be permanently reinforced with concrete pavement. All of this construction was planned to progress gradually but steadily as increasing property values warranted. The plan was thoughtful, methodical, and cautious—like Eaton himself. It became the backbone of Los Angeles flood control for the rest of the century and represented the apex of the local flood-control technocracy's accomplishments.

The engineering achievements of the Comprehensive Plan did not, however, stem the influence of politics and economics over the flow of water. Although based on heaps of expertly analyzed technical data that made the plan a model of technocratic problem solving, the program continued to encounter the kinds of nontechnical obstacles that had buffeted the Flood Control District since its inception. In particular, the plan's reliance on check dams as the primary means of debris control resulted from a combination of scientific understanding (and misunderstanding), old technical disputes among southern California engineers, the fallout of the San Gabriel Dam controversy, and the economic pressures of the Great Depression. Despite some engineers' reservations about the efficacy of these six- to ten-foot rock and wire structures, the Flood Control District under Eaton filled many canyons with them in order to control debris.[17] Although they were an appealing debris-control device under the historical circumstances in which the flood controllers worked in the early 1930s, they also left the county unprepared for the events of New Year's Eve 1933–1934.

Check dams originated in Europe and Japan, where they had been used to reduce the slope of steep canyons and thereby decrease the velocity of runoff. As water and silt collected behind the dams, vegetation grew, further decreasing the erosion and runoff from the canyon. Frank Olmsted, a member of the Board of Engineers appointed after the 1914 flood, studied the foreign examples and constructed an experimental set of check dams in Haines Canyon in 1915, with good results. Thereafter, however, problems arose. From the beginning, Reagan had sparred with Olmsted over the desirability of the barriers, contending that they would fail under heavy flows and merely divert the water in a new direction. In 1919, a landowner sued the district when check dam failures damaged his property, after which the district began holding the structures together with wire in order to make them more substantial. Eaton himself was ambivalent about them. His chief design engineer had reported to him in 1930 that the check dams were temporary, costly, and of "doubtful stability."[18] "Haphazard check dam building," Eaton cautioned in the Comprehensive Plan, "is inadvisable." He nevertheless concluded that when carefully implemented, the barriers could "constitute a most important emergency and flood protective measure." The plan requested a million-dollar fund for the district to build check dams as needed.[19] In the years after the completion of the Comprehensive Plan, hundreds of check dams were built in southern California canyons, including those above the La Cañada Valley.

Despite the structures' many drawbacks, the check dam program made sense, given the incomplete information on erosion in the San Gabriel Mountains. The San Gabriels wear down faster than almost any other mountain chain in the world, so engineers had little precedent to guide them. In fact, in the 1910s, flood-control engineers had not considered the debris flows to be a problem at all. With few people living in the foothills, the word *debris* hardly appeared in engineering reports at all; nor did it figure prominently in the minutes of the Flood Control Association meetings.[20] The main concern lay in controlling water on the more populated plain. Olmsted experimented with check dams only as a means of limiting the amount of water flowing out of the mountains.[21] The debris itself was not of much concern.

Even a 1928 debris flow that damaged twenty-five homes in Brand Canyon, above the city of Glendale, did not fully alert officials to the threat.[22] That flow deposited thirty thousand cubic yards of debris per square mile of watershed, less than a third of what future flows would yield. After the incident, the Flood Control District and the city of Glendale constructed a series of check dams in the canyon, again with good results. To the extent they were aware of the debris problem, however, the authors of the Comprehensive Plan underestimated the spatial extent of the danger. They were concerned about protecting the Glendale foothills that had been hit once before, and they realized that unchecked debris high in the mountains could quickly silt up the small mountain reservoirs they intended to build. They did not, however, imagine the extent of destruction that could occur over a space as large as the entire La Cañada Valley. Neither Montrose nor any of the other communities in the valley, for example, were mentioned on the plan's list of the county's most flood-threatened locations.[23] Given the underestimation of the debris threat and the apparently successful experiments in Haines and Brand canyons, check dams appeared to many people to be a sufficient solution.

Political conditions of the 1920s and 1930s enhanced the appeal of check dams. According to the engineer J. B. Lippincott, there was a "popular" demand for check dams in the Flood Control District's early years, and some were built under Reagan, largely against his wishes.[24] In his 1926 flood-control bond report, Reagan had advocated additional types of protection for La Cañada Valley, but when that measure failed and he resigned, the strongest opponent of the dams left the scene. Meanwhile, public pressure for check dams mounted. On a rainy day in January 1930, the Municipal League held a flood-control symposium.[25] Although the rains threatened to wash out the meeting entirely, the league managed to assemble a crowd of city officials, citizens, and flood-control engineers. At the meeting, Reagan's old nemesis Frank

Olmsted extolled the virtues of check dams and touted his success in Haines Canyon as proof. The district had not relied on them even more, Olmsted insinuated, because unlike the "big dams which were the central feature of the Reagan program," check-dam building offered no one "a big reputation."[26] As a result of the meeting, the Municipal League adopted a resolution calling check dams an essential part of flood control in the county and urging the Board of Supervisors to appropriate money for their construction.[27]

The political climate favorable to check dams was closely related to economic difficulties. From the mid-1920s onward, the Flood Control District grew increasingly unpopular at the same time that its funding sources dried up. First, the voters defeated the 1926 bonds. Then after 1929, with the onset of the Great Depression and the aftermath of the San Gabriel Dam scandal, the Board of Supervisors hesitated to place a new bond measure on the ballot to finance the thirty-three-million-dollar Comprehensive Plan. Meanwhile, section 15 of the 1915 Flood Control Act, which the California Supreme Court upheld in the 1931 case, required the Flood Control District to spend the twenty-two million dollars left over from the 1924 bonds on a dam in San Gabriel Canyon or some equivalent project.[28] Finally, the California government, which had been contributing three hundred thousand dollars annually to the Flood Control District's budget, withdrew this funding in 1933, as the state's economy staggered and the old north-south intrastate rivalry flared up again.[29] In 1931, Eaton estimated that at the current rate of funding, it would take twenty years to complete just the "Immediately Needed Projects" of the Comprehensive Plan.[30]

Amid these financial woes, check dams were an inexpensive strategy that added value in small increments. Although their brief life spans made them expensive over the long haul, check dams were cheap in the short term and an attractive option when public funds were scarce.[31] In contrast to large dams, which were valuable protective devices only once large amounts of money had been spent, check dams were individually inexpensive—about twenty-five to three hundred dollars apiece, depending on the size—and each one offered a slight increase in flood protection. A few check dams in a canyon promised at least slightly improved flood conditions until a full complement could be built. Finally, check dam construction was low-tech and labor-intensive, thereby offering an excellent opportunity to employ inexperienced, unemployed workers from the Depression-era relief rolls, who could construct the barriers with minimal supervision.[32]

Because funding obstacles prevented the district from doing much else in the way of new construction in the early 1930s, check dams constituted an important part of the agency's program. In item 22 of the Comprehensive Plan's most urgently needed works, Eaton asked the Board of Supervisors to approve a discretionary fund to be spent as needed on check dams.[33] For the next few years, check dams constituted virtually the Flood Control District's entire foothill debris-management program. As the officials put off the more expensive permanent items in the Comprehensive Plan, they built hundreds of check dams in Pickens Canyon and in other places upstream of the foothill debris cones. For the most part, the structures worked as planned. Water and debris backed up behind them. Out of the debris sprouted vegetation, adding to the mass of material situated in the canyons. No storm large enough to scour out this material passed during the early 1930s, so the debris kept piling up. Thus, these dams, which were the products of limited geological knowledge and the imperatives of political and economic conditions, came to influence how water flowed in Los Angeles County.

Like the check dams, Eaton's Comprehensive Plan as a whole reflected a mix of both engineering concerns and political economy considerations. Despite the significant engineering achievement that the plan represented, unrelated factors such as the falling credibility of the Flood Control District and the onset of national economic depression coincided to make some elements of the program more workable than others in the early 1930s. With funding obstacles blocking the plan's implementation, the district applied, in September 1933, for money from the Public Works Administration (PWA), one of the many federal agencies handing out money under the auspices of the new president, Franklin Roosevelt. As 1933 drew to a close, with the first dam in San Gabriel Canyon completed, the check dams proceeding, and the possibility that PWA money might soon be available to launch the rest of the Comprehensive Plan, water seemed to be flowing as planned.

PERILOUS EDEN

The New Year's Eve flood was a new type of disaster that stemmed from old phenomena. Debris had been flowing out of the foothills as long as there had been mountains in southern California. In fact, debris from the mountains had created the geography of the basin. But a few decades of urban growth remade the face of that landscape, changing both how water flowed and what was in its path. While flood planners ran in place during the 1920s, the potential for flood damage escalated. Major debris flows, however, happen at long intervals, and while all this growth took place, the mountains were relatively quiet. Into this changed context in 1934 poured the first terrific debris flow in decades. Like the flood of 1914, the New Year's Eve debris flows were normal events made disastrous by the particular set of circumstances in which they struck. Unremarkable events happening on geological, historical, and daily timescales came together in an explosive way and with destructive results.

The first ingredient that makes a debris flow is mountains. The steep slopes of the San Gabriel Mountains and other, smaller ranges rise out of the coastal plain to pinnacles that crown the Los Angeles Basin. Peaks ten thousand feet and higher stand only fifty miles from the sea. From base to tip, the San Gabriels are taller than the Rockies, and they are rising at the same rate as the Himalayas. Tectonic movements are forcing them upward. But they are coming down almost as fast. The faulted blocks of the San Gabriels push and grind against one another, pulverizing the rock in the process. Even to a layperson's ears, the terms geologists use to describe the rock indicate its unstable character: deformed, decayed, disintegrated, decomposed, rotted. By *deformed,* geologists mean that the fault blocks are internally broken, fractured, and easily separated. There are few large masses of solid rock in the range. By *decayed,* they mean that rock along the fault lines has been crushed into a clay so soft that, if you moisten it, you can mold it in your hands. Crumbled and crushed, pulverized and lifted, the exposed rock in the San Gabriels breaks off easily. It has, in combination with the steep slopes, produced a shallow, rocky soil mantle that erodes easily.[34] With weathering and tectonic movement, the mountains are coming down.

They are also burning up. Life is hard on the steep rocky slopes of the San Gabriels. Hot dry winds often blow from the desert side. Long rainless summers—and sometimes rainless winters—make water scarce. Fire is always a threat. The plants that eke out an existence there are highly specialized to the rugged conditions. The ceanothus, manzanita, chamise, and other hardy slope-dwellers grow in mosaics of dense patches. Collectively this brush is known as chaparral, a term that derives from the same Spanish word that inspired the name "chaps," for the thick leather leggings that herders in Mexico and the western United States wear to protect themselves as they ride through the thorny vegetation.[35] The chaparral crackles and crunches. Take a leaf in your hand. Press. Snap. It crumbles. There are jagged edges, spines, and prickles, all of which maximize the flammable surface area of the foliage. Into these tinderboxes, hot desert winds blow. Lightning strikes. Flames swallow the thickets and roar for more, sweeping up and down the slopes. The conflagrations, however, are not disasters. In fact, the plant community requires fire for regeneration. The heat melts the resin that seals seed pods. Fire fertilizes the soil. Without fire there would be no chaparral—and vice versa. The fire also does something else curious. It chemically alters the soil so that, just below the surface, the ground becomes impermeable to water.[36]

Water is the final ingredient. Streams carry sediment out of the mountains and deposit it on the gentler foothill slopes. Bit by bit, over thousands of years, this process built the bed on which the La Cañada Valley and other foothill areas lie. Mostly this was an incremental process, but occasionally it was violent. Storms can shower astounding amounts of rain on the mountains. When the rains come, especially when they fall on burned hillsides, water saturates the topsoil, and then the whole mass sloshes down the slope over the impermeable layer, gathering twigs, brush, stones, soil, ash, and any other debris that lies in its path. Sludge from various hillsides concentrates in the canyons. With enough runoff, it forms a viscous mass that lifts boulders and automobiles and even houses as it flows. Water may run over the top of the flow, or the debris may form barricades. Water piles up behind it, saturates the glob, and the whole mass slips farther along. Slowly the water drains from the mass until it is too dry to flow. Then it stops, waiting for more water. Often the globs will wait for decades until a storm like the New Year's Eve deluge comes along and starts them up again. In the meantime, they just sit. All of this makes it impossible to know when or where or how fast the masses will flow.[37] But when they do, they are irresistible, moving along at five to ten feet per second on the detritus cones, lifting or smashing anything in their path. The rolling and grinding of the boulders, one man observed, can be heard for miles.[38]

By the 1930s, one of the most dangerous places to live in southern California was the La Cañada Valley. Located at the northern limit of the city of Glendale, about twelve miles north of downtown Los Angeles, the valley rested on a sloping bed of alluvial deposits, three and one half miles long and two miles wide. The southeast end lay at 1,200 feet in elevation, and the valley rose up to 2,100 feet at the northwest corner. The Verdugo Hills flanked the southern edge of the valley. To the north towered the San Gabriel Mountains, climbing 3,000 feet in less than two miles to a pinnacle at Mt. Lukens, 5,050 feet above the Los Angeles Basin. Sixteen canyons were aimed at the valley, discharging water in well-defined channels. As it flowed onto the gentler inclines, the water spread and deposited its silt load. Much of the water percolated into the porous sediments on which the valley lay; the rest meandered over the cone in numerous wide but ill-defined channels with no banks to speak of. It was on this lower portion, less rocky and less steep, that developers first chose to build the densest residential subdivisions. But the low channel banks meant that water could run anywhere in the valley. Moreover, the entire valley was strewn with boulders, testifying to titanic floods past.[39] It was a dangerous geography.

The geography also made the foothills attractive places to settle. Among the people the foothills attracted was Harold Hedger. He had been born in 1898 about fifty miles east of Los Angeles in Riverside, which was then covered with orchards, and he moved to Long Beach just as that city began to thrive during the final years of the harbor construction. His father had a wagon and team and earned a living hauling goods. As all of southern California boomed in the late teens, Hedger was among the many to grow from an agricultural and small-commercial childhood into an urban professional adulthood. After his sophomore year in college at the University of California, Berkeley, he joined the navy for two years. Upon returning, he caught on with the Los Angeles County Road Department, then embarking on its busiest decade of road building yet, as it paved the way for automobiles to replace the horses and wagons of people like Hedger's father. Hedger resumed his studies at Berkeley, where he wrote a thesis on San Gabriel River flood control. During the summers, he worked for the Flood Control District under Reagan. As a surveyor, Hedger spent considerable time in the field, living in mountain tent camps with teams of other district employees.[40] At their camp in San Gabriel Canyon, the workers trapped animals. One man even shot a mountain lion one morning. "There was all kinds of wild game back in these hills," he remembered, "and the minute you got away from town, you were right back to nature."[41] By the end of the 1920s, however, the mountains were not so remote. Their foothills were suburbanizing as people flocked to the heights to escape the heat of the plains and the din of the city. Hedger himself, now married, moved with his wife to the foothill suburb of Montrose.

The Hedgers' move was part of a larger migration into the foothills that eliminated their remoteness and created the La Cañada Valley communities of La Cañada, Montrose, and La Crescenta.[42] There is some uncertainty about the origin of the name "La Crescenta," but some historians have attributed it to the Spanish word *la creciente,* meaning "the flood tide," in reference to the water that occasionally gushed out of the canyons. Historically part of a Spanish land grant, these highlands attracted a few nineteenth-century health seekers, who occupied quarter-mile-wide hillside tracts, but as late as 1908, only 150 people lived in the valley.[43] That began to change in the 1910s. In 1913, Frank Putnam Flint subdivided Flintridge; E. T. Earl subdivided a five-hundred-acre ranch, creating Alta Canyada; land sales in the community of Montrose opened; and electric streetlights illuminated the valley for the first time. In 1915, the state paved Michigan Avenue, the valley's main thoroughfare, and added it to the state highway system, renaming it Foothill Boulevard. Over the next decade and a half, the most rapid growth of the valley occurred. A local newspaper, the *Ledger,* began publication in 1922; more communities and neighborhoods were subdivided; the Southern California Gas Company installed lines in 1923–1924; telephone service arrived in 1925; and electric streetcars began operation in 1930. From the 150 souls spread out on large tracts in 1908, the valley had grown into a suburban area with several thousand residents, paved streets, and all the modern amenities of 1920s city life. The *Los Angeles Times* called the La Cañada Valley "a region of crystal-clear air, wonderful vistas, delightful homes—a real Southland place of beauty."[44] All of this growth occurred on alluvial beds laid by past floods.[45]

As had been the case before the 1914 flood, however, most of this danger was invisible. Hedger chose high ground for his home, but few were as savvy. Developers, sometimes unscrupulous and more often ignorant themselves, sold homes in flood-prone spots to unsuspecting home buyers during the real-estate boom of the late teens and twenties. And why not? Los Angeles was the land of sunshine. Who would worry about floods? The decade's rapid population growth, however, coincided with what Eaton termed an "unparalleled dry period" that lasted from 1916 to the early 1930s. That meant that as in 1914, southern Californians had little collective memory of floods. Technically, real-estate developers had to file information about a site's flood history, but this was widely ignored during the dry years, and few home buyers thought to check anyway. Only 8 percent of them, Eaton estimated, had experienced the deluge of 1914. The public ignorance was aggravated by legal restrictions, which forbade the Flood Control District from spending money to publish reports or other documents for public distribution, except in connection with an election.[46] Consequently, citizens had to rely on information reported by the media, which invariably focused on political scandals and water shortages. Even the best-informed citizens feared too little water more than too much. They thought of debris control as something that took place far away, high up in the San Gabriels. More than anything, by the 1930s, citizens associated flood control with scandal, graft, and collapsing dam sites. Few knew that it could also be about mud in living rooms.

Then, in late 1933, fire returned to this perilous Eden. Although a historically frequent visitor to the foothill ecosystem, fire had been absent from the watershed above the valley since 1878, well before the beginning of urbanization. A chaparral fire broke out in Pickens Canyon on the edge of the Angeles National Forest around eight o'clock on the evening of 21 November, burning decades of accumulated fuel. The next day, firefighting crews seemed about to bring the blaze under control, but dry Santa Ana winds blew in from the desert, fanning the flames. For the next three days the inferno raged, spreading into the adjacent canyons. Finally the flames halted at a mountain road in the Arroyo Seco Canyon. When the embers died out, a seven-and-a-half-square-mile swath of moutainside above the La Cañada Valley lay barren. Seven houses in Briggs Terrace, a 250-home subdivision under construction, had burned to the ground. Fearing severe erosion, the U.S. Forest Service sowed the entire area with red mustard, but the seeds never had the chance to germinate.[47]

Only a few weeks later, the New Year's Eve rains drenched the canyons and saturated the soil. For several years the check dams had trapped debris that formerly would have trickled out of the canyons during light storms. That night, every check dam failed—and there were hundreds of them. When they crumbled, the material they had collected and the vegetation that had grown over them joined the detritus from the fire in sloshing down the canyons.[48] Because of the check dams and the long floodless stretch, the channels on the upper reaches of the floodplain were deeper than in the past. As a result, they carried the debris farther down the slopes and closer to the development. Thus, the opportunity was lost for the torrents to spread over the upper reaches of the debris cone and dissipate their energy there.[49] Instead, the debris crashed with full fury farther down the cone, depositing a fifty-nine-ton boulder right on Foothill Boulevard. There the masses spread, crossing the road into the heavily residential sections of the valley, where the stream banks were low and channels undefined. Streets conveyed the debris into buildings and onto lawns, coating a quarter of the valley with rubble and spreading destruction everywhere. By the time the flows reached the

Flood Control District's defenses on Verdugo Creek below Montrose, they had already done five million dollars in damage and taken forty lives.[50]

The 1934 disaster differed in many ways from the 1914 inundation. One involved water on the plains; the other debris in the foothills. One did damage countywide; the other was mostly confined to seven square miles. Despite these differences, however, they stemmed from strikingly similar circumstances. Both events, although triggered by sudden storms, had roots that lay deep in the dynamic ecology and the growth-oriented economy of the region. In both cases urbanization had placed people and development in the path of flowing water, and in both cases, that urbanization also changed how and where that water flowed. Both storms struck after an extended floodless stretch, during which population growth and short human memories created populations oblivious to the danger. And in 1934, as in 1914, the water that fell transformed previously harmless and invisible changes to the urban ecosystem into catastrophe. Debris smothered the La Cañada Valley on New Year's Eve for substantially the same reasons that silt had choked the harbor in 1914.

AFTER THE MUD

As similar as the causes of the two floods were, however, the public responses to them could not have differed more. Whereas southern Californians formed the Board of Engineers and the Flood Control Association within weeks after the 1914 deluge, the New Year's Eve debris flow spurred little immediate action. It discredited the check dam strategy and prompted consideration of new alternatives, but the district found it difficult to marshal political and financial support for new programs. In May 1934, the PWA denied the district's application for loans submitted the previous September. More critically, fallout from the San Gabriel Dam fiasco still loomed over the district's efforts, tying its hands legally, politically, and fiscally. Not until September 1934 did the Flood Control District put forward a bond measure to raise money to protect the foothill communities. When the bond issue went before the voters in November, the San Gabriel Dam fiasco and the crisis of the Great Depression still dampened voter enthusiasm, and the measure failed. For the moment, little would be done by the district to avert a repeat of the New Year's tragedy. Thus, the flood did not catalyze political will the way the 1914 inundation had. Instead, the events of 1934, culminating in the defeat of the bonds, illustrated the limitations of local flood control as then constituted. The 1915 Flood Control Act and the bond issue of 1924 had aimed to prevent water from overflowing the plain. The legal, political, and financial arrangements that made up the flood-control assembly line, however, were too inflexible to be redirected to control debris in the foothills.

The debris flow taught a number of important lessons, though. The Conservation Association of Los Angeles County convened at the Biltmore Hotel in downtown Los Angeles on 23 March 1934. Founded in 1924 after a summer of devastating mountain wildfires, the Conservation Association was closely tied to the Los Angeles Chamber of Commerce, with which it shared an office, a staff, and several officers. The Conservation Association president was, ex officio, a member of the Chamber of Commerce Board of Directors, and the Conservation Association executive secretary served as manager of the Conservation Department of the Chamber of Commerce. The purpose of the organization was to promote fire prevention and water conservation and to give the chamber influence over public conservation education, technical advice to governments, and the shaping of conservation policies. After the New Year's flood, the association's president, William Rosecrans, assembled a number of flood and forestry authorities to discuss the debris threat. The flood, they agreed, had revealed the precarious existence of foothill residential districts, particularly those below fire-denuded slopes.[51] It also discredited check dams as a method of protecting against the menace. "It is now quite evident," declared the engineer J. B. Lippincott, "that check dams are a failure."[52] The flood, however, did not just discredit the old strategy; it also suggested a new one—debris basins—almost by accident.

The canyon just west of the La Cañada Valley was known as Haines Canyon. Until recently, the mouth of Haines Canyon had been the site of gravel mining operations, which had left a huge pit at the base of the mountains. Residential development supplanted the gravel mining, and the town of Tujunga incorporated in 1925. Eventually, the Flood Control District consolidated the abandoned gravel mining site into its flood-control works. More than a hundred feet deep and several hundred feet across, the pit was a formidable obstacle between the canyon and the town. On New Year's Eve, the check dams in Haines Canyon failed, just as they did just over the hill above the La Cañada Valley. But instead of enveloping the new town, the ten-foot-high waves of debris that stormed out of Haines Canyon slumped into the old gravel pit. Clear water streamed out the downhill side, sparing the town of Tujunga. Meanwhile, another converted gravel pit, this one on the Verdugo Wash downstream from Montrose, caught ninety thousand cubic yards of debris, saving Glendale from a fate similar to what befell the La Cañada Valley.[53]

When Eaton spoke at the Biltmore in March, he touted these pits as the solution to the foothill debris threat. Debris basins, as the pits were called, consisted simply of a hole in the ground with a drain or spillway on the downstream side. If strategically placed at the canyon mouths, the basins would trap flood debris and allow water to drain out harmlessly. The concept was not totally new. The U.S. government had used them with success in Utah. Until the 1934 flood, when the basins proved their usefulness in Los Angeles, however, the district's reliance on them was minimal. Reagan had advised building a few debris basins in his report recommending the 1926 bonds, which the electorate defeated, and the Comprehensive Plan of 1931 had also called for a number of debris basins, including several above the La Cañada Valley. They were not considered among the district's highest priorities, however, and none had been built by 1934. The PWA application the previous September had requested money for only two. Despite hundreds of locations countywide that might have benefited from a debris basin, there were only a handful of such basins in operation in the district, and some of those, like the one in Haines Canyon, were the result of accident as much as deliberate planning. The basins had to be cleaned out after each storm, of course, but the twenty-cent cost per cubic yard of removal was only one-fifth what it cost to remove debris from streets and yards. Inspired by the success of the basins in Haines Canyon, the Flood Control District officials envisioned building a Maginot line of debris basins along the front of the foothills.[54]

The old financial difficulties lingered, however. The Flood Control District had gained no new bond money since 1924, and the Flood Control Act's section 15 prohibited the diversion of the 1924 bonds to meet the new debris threat. The district's last hope evaporated in May 1934, when the PWA rejected the application for a federal loan. Although PWA reviewers concluded that "the comprehensive plan of the District, in general, appears to be a logical solution of the flood problem," they feared that the district would have difficulty levying taxes to pay off the loans because of "inherent defects in the 1915 Flood Control Act."[55] The only way to get money to construct protection for the foothill communities, which awaited only another rainstorm before an even worse disaster might befall them, was to go back to the voters with another bond proposal.[56] The San Gabriel Dam fallout threw one more obstacle in the way in September 1934, when Eaton, recently indicted by the grand jury for his role in the scandal, resigned in disgrace. On that note, the Flood Control District began its campaign for more money from the voters.

Given the economic conditions of the 1930s and the Flood Control District's checkered history, the voters of Los Angeles County were in no mood to give the agency another round of financing. In hopes that the fresh memory of the New Year's Eve devastation would soften voters' sentiments, however, the district persisted. The interim chief engineer, Samuel M. Fisher, prepared a report based on the Comprehensive Plan and the newly recognized need for debris basins.[57] Of the twenty-six million dollars it requested, four million would go to a set of debris basins and other works to protect the foothill communities. The rest was to be allocated to other projects scattered around the district in hopes of luring votes from people outside the beleaguered foothills. As in 1924, the bond proposals were designed by engineers and presented to the electorate at the last minute, with little opportunity for public debate or review.

As supporters gathered on Monday, 8 October, to hash out a campaign strategy, they faced strong opposition. The *Los Angeles Times* captured the sentiment in opposition to the Flood Control District. In this time of severe depression, the *Times* observed, "it is not now necessary to gamble $26,000,000" on a troubled and wasteful agency. Millions of dollars had been squandered so far, and "the prospects for further losses in the same program seem inescapable." The *Times* also chided the Flood Control District for doing nothing all year long on the issue and charged that the plans currently before the voters were inadequately prepared.[58] Making the bonds even more unpopular was the announcement on 9 October that San Gabriel Dam No. 1, the larger of the two dams in the revamped San Gabriel project, would cost five million to six million dollars more than the initial ten million dollars estimated.[59] Fisher requested a halt in construction to revise the design.[60] "There has been some bad engineering somewhere," the *Times* seethed; taxpayers could not be asked to keep paying for "a piled-up series of costly errors."[61] With the memory of the San Gabriel Dam fiasco still fresh, a national fiscal crisis sapping state and local funds, and hints of yet more incompetence at the Flood Control District, not even another round of debris flows and flooding that did $150,000 of damage in a single day of cloudbursts three weeks before the election could sway the electorate.[62] Voters narrowly defeated the flood-control bonds by 26,000 votes out of the total 675,000 cast.[63]

The defeated bond measure of 1934 represented the last attempt by the Flood Control District to direct flood control on its own. The district would live for another fifty years before being absorbed in the County Department of Public Works, and its Comprehensive Plan (amended to include debris basins) would form the basis of future flood control. After 1934, however, the Flood Control District relied on the federal government to supply the money and to oversee the design and construction of most flood-control projects. The events of 1934 were a fitting conclusion to two decades of local flood control. The year began with a flood disaster made by a combination of earth, fire, rain, and urbanization, and it ended with politics and economics thwarting efforts to prevent repetition of the event. The flood itself suggested a new engineering plan, but the local flood-control regime proved incapable of carrying it out. The Flood Control Act's tax provisions prevented it from receiving a federal loan. A state supreme court ruling disallowed transfer of existing funds to new projects. Finally, political scandal and nationwide economic depression dissuaded the voters from supplying the money. As had been the case ever since the county first undertook an organized flood-control program, law, economics, and politics had as much to do with how water flowed in southern California as engineering did.

THE UNUSUAL HAS HAPPENED

When he addressed the postflood convention at the Biltmore in 1934, the Conservation Association's former president B. R. Holloway summed up the thinking that prevailed among flood-control advocates: "The unusual has happened," he declared.[64] As the La Cañada Valley and other vulnerable locales faced the 1934–1935 rainy season with no more protection than they had had during the disastrous previous year, flood-control leaders groped for explanations. Essentially they were asking why, after fifty-five million dollars and two decades of flood control, nature was still able to move more earth in twenty minutes than the Los Angeles County Flood Control District had moved in five and a half years in San Gabriel Canyon.

Like Holloway, most observers focused on the unusual. From the egregious mismanagement in the San Gabriel Dam fiasco, to the violent foothill fires and floods, to the rapid development of the metropolitan region—all signs led observers to conclude that local flood control was overmatched by extraordinary external conditions. This focus on the exceptional diverted flood controllers' attention away from the problematic foundations on which the flood-control program was built. Few people considered the ways in which the structure of flood-control law and politics contributed to the district's problems, nor did many observers question whether technology could control nature's fury. And almost no one challenged the proposition that developing the floodplains was desirable. Instead, the assumptions that had motivated and shaped the flood-control program thus far went unexamined. Consequently, the explanations for the failure of flood control led to a search for more powerful agents—more capable leadership, better technology, and more money—to accomplish the old goals of reengineering the rivers.

One set of explanations revolved around the extraordinary mismanagement, especially the corruption, in the San Gabriel Dam affair. In the 1934 election, Los Angeles County voters not only rejected the district's record by disapproving the bonds, but also elected a new Board of Supervisors whose members were committed to municipal government reform more generally. Two of the five seats changed hands, and a third member who had been appointed to fill a vacant seat at the end of the previous term was reelected. Supervisor John Anson Ford, a former midwestern journalist steeped in Wisconsin Progressivism, who would go on to be recognized as a champion of honest government, epitomized the new breed.[65] With flood control in mind, he lamented shortly after his election, "Again and again we have seen Los Angeles County sold on a good idea and then 'sold out.'"[66]

In the early 1930s, much of Los Angeles County agreed with Ford. The old criticisms of Reagan resurfaced, of course.[67] And one engineer described the district's policy so far as "opportunistic and drifting."[68] Even the methodical Eaton was not immune to the criticism. The Municipal League and the San Gabriel Valley Associated Chambers of Commerce had sparred over the location of San Gabriel Dam in the 1920s, but the grand jury indictments united them in a "Remove Eaton" campaign, which began in November 1933 and finally bore fruit when he resigned the following September.[69] During the 1934 bond campaign, the *Los Angeles Times* wrote that "the past history of flood control in this county has been—to put it mildly—unfortunate . . . a good many million dollars of the previous flood-control bond issue have been wasted."[70]

This focus on mismanagement of the Flood Control District led to a replacement of its managers. Echoing the Progressive Era thinking of twenty years earlier, Ford said the district's "first need" was "able, experienced leadership, untrammeled by political or vested interests."[71] In 1935, a variety of groups tried to remove flood-control authority from the Board of Supervisors. Two such schemes made it as far as the state legislature but no further.[72] Although the board resisted such moves, Ford and the other new supervisor, Herbert C. Legg, did lead the first steps toward placing Flood Control District employees under the civil service, in hopes that "politics would be eliminated from flood control employment."[73] Furthermore, the board searched outside the Flood Control District for a new chief, quickly settling on the Bureau of Reclamation's crusty, no-nonsense engineer Cleves H. Howell.[74] With Howell's appointment, Ford said, "the whole flood control situation is in new hands and its policies and technical decisions will henceforth be determined by objectives of economy and popular welfare."[75] Thus, by the end of 1935, the district had reorganized but not reoriented. Ford and others, focusing on the gross corruption and mismanagement of the previous decade, believed that they could address the problems of the district by changing the managers. But ultimately, even though different people were making the decisions, the technocratic goals and policy-making methods were essentially the same as those of the old regime.

Another set of explanations for the failure of flood control blamed extraordinary natural events. "Normally," Eaton wrote, "mountain and foot-hill watersheds are covered with a chaparral type of vegetation, efficiently functioning as retarders of flood flows and controlling erosion."[76] The implication was that the fire and rain that had triggered the debris flows were abnormal. Eaton and others acknowledged that such events were clearly within the realm of what the Los Angeles region was capable of producing, but at the same time they focused on the unusual aspects of the events. The watershed had not burned since 1878. The record showed few storms that matched the intensity of the New Year's Eve downpour. The scientists gathered for the Conservation Association's Biltmore conference "unanimously emphasize[d] the astonishing increase in flood water and detritus from a burned watershed as compared to an unburned area under similar conditions."[77] The California Forest Experiment Station's senior forester presented tables showing that during the New Year's Eve storm, the denuded Pickens Canyon yielded forty times as much runoff and a thousand times as much erosion as unburned watersheds.[78] He concluded that "the cause of floods in southern California lies in neither fire nor rain alone, but in the sequence of severe fire followed by intense and heavy rainfall."[79] Severe, astonishing, abnormal—these explanations were accurate, but they overlooked the fact that the allegedly normal chaparral cover that would have prevented erosion on unburned slopes was actually a cause of fire and that rainfall was a cause of chaparral. In a sense, fire and flood were not extraordinary, but rather were very normal parts of the creation and evolution of the foothills' and forests' ecology. By casting the ordinary as extraordinary, the explanations deterred public action. As James Jobes of the Army Corps commented, the 1934 disaster did not fully impress the public, because the combination of the fire and flood were considered "a coincidence, the probability of recurrence of which was very remote."[80]

Instead, that perceived coincidence brought pleas for high-tech firefighting. The county surveyor, for example, called for better-trained and better-equipped firefighters to operate a system that would enable them to reach any fire in the mountains with no more than five hundred feet of hose. He proposed pock-marking the hills with central reservoirs and connecting them with steel pipes to secondary reservoirs and hydrants that would be strategically located. The hydrants would be specially fitted to connect to either U.S. Forest Service or Los Angeles County hoses. The pipes were to be Grade A steel, mill-tested to eleven hundred pounds of pressure. The steel pipes, hydrants, and networks of reservoirs to keep water flowing composed a mechanistic system designed to eliminate fire from the watersheds. Indeed, William V. Mendenhall, the supervisor of the Angeles National Forest, promised that with adequate facilities, he would almost guarantee the suppression of any fire with less than one acre of watershed lost.[81] Believing that an unusual sequence of fire and rain was the problem, county leaders schemed to eliminate such extraordinary coincidences and to perpetuate the allegedly normal conditions in which fire was absent and vegetation slowed runoff and erosion.

A final set of causes to which observers ascribed the shortcomings of local flood control was the astounding rate of growth that Los Angeles had experienced in recent decades. Between 1914 and 1930, the population rose 180 percent, and the total assessed valuation in the county rose from $850 million to $4 billion. This urban development, flood controllers agreed, had proceeded much faster than the construction of new protective works. Eaton, for example, introduced his 1935 article on flood control by citing recent Los Angeles growth statistics and the challenge they posed to flood controllers. The prominent civil engineering journal *Engineering News-Record* blamed "phenomenal growth" for making the flood problem "more acute." All of this growth left 380,000 people and three hundred million dollars of property vulnerable to flooding. With such development, Jobes said, "the flood control engineer is hard pressed to keep ahead of it."[82] None of this, of course, should have been surprising. Growth, after all, was what almost everything in Los Angeles society, including flood control, was designed to achieve. Still, flood controllers seemed to treat it as an external force, as much beyond their control as the rain itself.

Although southern Californians offered a variety of explanations for the county's inability to control the flow of water, all these explanations focused on the extraordinary external causes, an approach that had the effect of intensifying, not changing, what was already being done. They believed that Flood Control District mismanagement problems could be solved by changing the managers without reexamining what exactly it was that they were managing. They also assumed that the extraordinary power of fire and flood could be resolved with a better firefighting machine to go along with the flood-fighting machine. Similarly, the recognition that rapid urban growth aggravated Los Angeles's flood problem led to the desire for building even bigger and better engineering structures to accommodate such development. Collectively, the explanations did not question the assumption that humans could reengineer the rivers to make them safe or the structure of flood control that vested power in experts to execute this goal. They did not challenge the idea that the region's potential for growth was unlimited or the idea that any flood-control measures in the service of that growth were justifiable. Rather, the explanations testified that the resources thus far used were insufficient. As the Army Corps's colonel Arthur H. Frye later reflected, the floods "proved beyond doubt that the problem was too great to be handled by local government."[83]

FAILURE IN THE URBAN ECOSYSTEM

These explanations were not wrong. Corruption and mismanagement *had* thwarted the San Gabriel Dam project. Nature *had* done something very powerful and unexpected in 1934. Development *had* outstripped the advance of flood-control works. And indeed, the whole problem was very complex. But these explanations, by focusing on what seemed extraordinary, missed a crucial element of the problem: the interaction of ordinary elements of the urban ecosystem. In building a city in the path of floods, southern Californians narrowed the number of possible ways that water could flow without being destructive. At the same time, structures such as paved streets and even check dams increased the number of ways that things could go wrong to make water flow destructively. Finally, urban law, politics, and economy interacted in ways that prevented flood-control solutions from working. The city's growth multiplied the ways for disaster to happen at the same time that it reduced society's tolerance for such events.[84]

Los Angeles County flood-control planners, however, were so committed to extraordinary explanations and held such great faith in engineering structures that they could not hear the few voices that focused on ordinary aspects of the flood problem. Donald Baker, a prominent southern California engineer, bucked the prevailing technocratic wisdom by suggesting that Los Angeles flood control was "preeminently a problem of economic and human nature, with construction and hydraulics secondary."[85] In a 1935 roundtable discussion published in the *Transactions of the American Society of Civil Engineers,* Baker contended that the "physical, meteorological, hydrological, recreational, cultural, historical, financial, and political" aspects of flood control all contributed to the problem. He called for a comprehensive plan in which "all the various phases and aspects of the problem" would be coordinated. Until Los Angeles County adopted such an approach, he wrote, floods would recur.[86] But Baker's proposition that managing water also meant managing politics, economy, and society was simply too foreign even to be heard. In dismissing Baker's comment, Eaton wrote in the same roundtable that the county had already produced *the* Comprehensive Plan.[87] So different were the premises behind Baker's comprehensive plan and Eaton's comprehensive plan that they misunderstood one another despite using the same words.

Baker provided few specifics as to what his comprehensive plan would look like, but by the mid-1930s, it was clear that Los Angeles flood control needed to try something new. Up to that point, the district had made little progress. It had spent nearly sixty million dollars, mostly on mountain dams, which provided no defense against floods of the type that overflowed La Cañada Valley. Of the nearly 500 miles of river channels within the district's jurisdiction, 273 remained unimproved, and 184 more had only temporary protection measures.[88] As temporary measures, the district had piled the levees higher and dug the beds deeper. Between those confines, the water sometimes filled a narrow channel and clipped along at a considerable pace; other times it divided into several channels or just puddled in stagnant pools. There were boulders and willows and other obstacles and, in dry seasons, lots of dust. In some places, engineers had armored the sides of channels with wooden pilings, rocks held in place by wire, or even old automobiles, in hopes of forcing the rivers to stay in one place; elsewhere, they had layered the beds and sides of levees with chunks of concrete several feet in diameter, called riprap. But these reinforcements had already outlasted their intended life spans, and they had rotted, rusted, and washed away over the years, their maintenance neglected for lack of funds.[89]

Despite the scant progress of the previous twenty years, the flood-control outlook was perhaps not so bleak. By 1935, the Flood Control District had formed a partnership with the U.S. Army Corps of Engineers, and over the next three decades, the two would spend nearly two billion dollars to erect more substantial protections along the rivers, an endeavor that produced one of the largest public works projects in world history and remade the hydraulic landscape of Los Angeles.

CHAPTER 5

The Sun Is Shining over Southern California

The Politics of Federal Flood Control in Los Angeles, 1935–1969

It was an austere landscape. Standing in the bed of the Los Angeles River in the late 1960s, you would have found yourself in a field of concrete. A smooth veneer of pavement covered the trapezoidal channel for miles straight up- and downstream and from side to side, hundreds of feet across the bed. Also confined in pavement were the levee sides that angled upward toward the sky. The scene resembled nothing so much as an empty freeway. At the center of the bed, water flowed in a groove several inches deep and several feet wide. This narrow and shallow low-flow channel, as it was called, kept water moving year round at a velocity fast enough to prevent the current from spreading out, slowing down, and depositing silt in channels. There were no plants, no rocks, no mud, no dust, no curves, just sun glinting off white pavement as far as the eye could see, a scene broken only by the blue ribbon of water trickling in the low-flow channel and the freeway overpasses flying overhead. It was clean, smooth, white, geometrical—a picture of hydraulic order.

Reflecting in 1965 on the remarkable transformation that had taken place in the riverbeds of southern California, Harold Hedger lauded the Army Corps. Without the corps, he speculated, the job would have taken twice as long as it did.[1] Many subsequent observers have come to similar conclusions, identifying the involvement of the corps as the turning point of Los Angeles flood-control history, the point at which all the false starts and failed plans finally matured into systematic flood protection for the Los Angeles Basin.[2] Without a doubt, the Army Corps of Engineers brought considerable technical expertise and badly needed federal money to bear on a problem that had confounded local efforts for more than two decades, but the emphasis on the corps's contributions is something of a deus ex machina, the engineering version of the cavalry inevitably riding to the rescue at the end of some 1950s Western movie. That emphasis misses the deeper questions of why the corps got involved at all and what enabled the federal engineers to have so much apparent success in taming the waters. In short, praising the technical wizardry of the Army Corps is accurate but incomplete; it obscures the many contingent factors of politics and climate that limited the range of options that could be imagined and implemented and that facilitated the engineers' success. These questions are important because, while Hedger and others exuded confidence in their handiwork in the 1960s, within a decade, many southern Californians, including some engineers, were doubting whether the accomplishments of the midcentury technocracy were triumphs at all.

PATHS NOT TAKEN

Hypothetically, at least, the flood-control failures of the early 1930s might have sparked a reevaluation of the program's technocratic structure and pointed flood controllers down a new path. In fact, during the Depression decade, a few tentative efforts to integrate less-technical alternatives into flood control did surface. Many conservationists advocated so-called upstream flood-control measures—fire prevention, reforestation, soil and water conservation, and check dam building. An urban planning firm proposed combining parks and recreation with flood control. And a few people even dared to challenge the sacred pursuit of growth by suggesting hazard zoning to limit development of the floodplains. None of these alternatives, however, got very far. In fact, most of them failed even to come up for serious public discussion. Instead, the structure and culture of flood-control politics in 1930s Los Angeles rendered these alternatives virtually invisible. Consequently, this crossroads decade ended with southern Californians reembarking with renewed determination on the same technocratic path they had been on since the 1914 flood. Only this time, they had a powerful federal companion in the U.S. Army Corps of Engineers.

The alternative that would have differed most from the previous flood control was hazard zoning. Through exercise of the police power, cities and counties in California had the authority to limit or prohibit development in flood-prone areas. Although more than two million people inhabited Los Angeles County in the 1930s, many of the areas that would later be threatened with flooding were still undeveloped at the time, and a 1942 State Planning Board report suggested that zoning to limit future development in floodplains would in many cases be more cost effective than erection of dams and channels to protect places from inundation.[3] It was widely recognized by the 1930s that development aggravated the flood danger—both by augmenting the runoff that storms produced and by placing more people and property in harm's way.[4] Moreover, flood-control works often created a sense of security that induced more development, which in turn necessitated more flood-control works. A few people in the 1930s sought to break this cycle with hazard zoning. The State Planning Board, the Municipal League, and an urban planning firm hired by the Los Angeles Chamber of Commerce all made little-noticed recommendations for using zoning to limit development in the flood plain and thus prevent flood damage, and Los Angeles County actually amended its zoning ordinance in 1940 to prohibit new residential construction in a pair of small flood-prone areas a few miles southwest of downtown.[5] None of these initiatives, however, developed into anything remotely resembling systematic hazard zoning. As one political scientist in the 1950s put it, "There has been some discussion as to the feasibility of zoning . . . but no action."[6]

This was due partly to the power structure that vested flood-control decision-making authority in technical bodies. The engineers, to whom flood control had been entrusted, were builders, not planners, and they defined the nontechnical aspects of flood control as outside the scope of their work. For example, in response to attempts to get the district to landscape its facilities and to open them to boating, fishing, swimming, and sight-seeing, the agency's officials maintained that it was neither their duty nor even legal to spend district money on such matters.[7] Similarly, federal engineers often chided city planners for failure to prevent development of flood-prone zones, but Congress did not authorize those engineers to assist local authorities with floodplain zoning until 1960.[8] Meanwhile, local engineers did not seem particularly interested in hazard zoning. Eaton, among others, believed that development had already proceeded too far to make land-use restrictions an effective option.[9] Thus although both federal and local engineers recognized that uncontrolled land use had aggravated the flood threat, they considered coordinated planning to be either unworkable or somebody else's responsibility. Consequently, any challenges to the engineering focus of Los Angeles flood control would have to come from outside the system. And as the San Gabriel Dam fiasco indicated, assembly-line flood control was not a system that was particularly able to cope with challenges from outsiders.

One such challenge, however, did emerge in 1930. In 1927, the Los Angeles Chamber of Commerce had commissioned the nationally renowned planning firms of Olmsted Brothers and Harland Bartholomew and Associates to develop a comprehensive public parks and recreation plan to address what was coming to be widely perceived as a near-crisis shortage of such amenities in the area. Although it was a private organization of southern California business leaders, the chamber had long exercised informal quasi-government powers. For example, it received annual appropriations from Los Angeles County to promote and advertise the region. It sponsored studies and conferences on municipal issues including traffic, flooding, and now parks.[10] And it acted as a liaison among government officials and agencies. All this, along with the tremendous wealth and political influence wielded by its membership, stamped the chamber's imprint on most municipal policies and gave it an unofficial source of power that its own parks report would shortly threaten. That 1930 report, *Parks, Playgrounds and Beaches for the Los Angeles Region,* outlined a $124 million proposal to use the county's land-use regulatory and landownership authority to meet the parks and recreation crisis. The planners observed that urban growth was destroying the very scenery and recreational opportunities that had made Los Angeles attractive in the first place.[11] They also contended that unfettered urban development along the rivers raised property values in flood-prone areas, making flood control both expensive and difficult. The planners sought to remedy both problems. Among their many recommendations, they urged the county to purchase swaths of land along the channels and landscape them to create a web of greenbelts that would be used to absorb swollen rivers during floods and devoted to recreation and auto touring and transport the rest of the time. Not only would this proposal provide more parks, the planners maintained; it would be a cheap, effective flood-control strategy. To the distress of many chamber leaders, however, the planners also proposed to create a new governmental authority that would have sweeping powers to raise money and purchase and develop property for parks, roads, flood control, and other infrastructure. The body would be appointed by the governor and would operate outside the "frequent political overturnings," over which the chamber wielded such influence. "Terrifying," one member of the chamber's Board of Directors described it. "Radical," said another. Apparently fearing that the new authority would infringe on their own informally exercised power, the chamber's directors reduced the print run of the report from seventy-five hundred to two hundred copies in an effort to prevent the rise of public support for the plan.[12]

The report also ran afoul of a political culture that worshipped growth and development. For decades southern California boosters had celebrated towns that did not yet exist, water that had not yet arrived, and people who did not yet live there. Much of the region's development had been driven by promises of future growth. The Olmsted and Bartholomew parks and flood-control plan not only proposed to enlarge municipal property ownership and create a new agency to manage it; it also advocated the use of hazard zoning to control riparian real-estate prices. Here the chamber's leaders balked at what they perceived as a radical threat to future property development in Los Angeles. Although by that time the city boasted some of the first and toughest zoning regulations in the country, these earlier initiatives had been intended to protect existing property interests.[13] To use zoning to limit future development was a leap that the business leaders of 1930s Los Angeles were not prepared to make. Some Chamber of Commerce directors were so uneasy about any sort of government interference with private property rights that they balked even at federal funding for traditional flood control, for fear of allowing "the Federal Government to become so intimate in our affairs." One director snorted, "We objected under the NRA to all of that, and now we are inviting them back in through another door."[14] Consequently, in contemplating the various flood-control alternatives in the 1930s, the chamber appears to have given zoning little, if any, attention.[15]

Thus a potentially effective flood-control tool became a path not taken. Even the authors of *Los Angeles: Preface to a Master Plan,* a book advocating planning for the city, failed to connect zoning and flood control. The flood-control section of the book barely mentioned zoning, and the zoning chapter completely ignored flood control.[16] Nor did the U.S. Department of Agriculture consider zoning a practical possibility. It devoted only three paragraphs to the subject at the end of a sixty-seven-page report on the Los Angeles Basin in 1941, concluding that "it would be very difficult to bring sufficient pressure to bear to initiate zoning regulations."[17] Despite the engineers' twenty-year inability to keep floods away from people, the possibility of using zoning or parks to keep people away from floods remained too contrary to the prevailing political structure and culture to receive extensive consideration.

Meanwhile, engineering remained a compelling flood-control strategy. President Franklin Roosevelt's New Deal launched an era of high-profile, large-scale federal engineering monuments that began with the completion of Hoover Dam in 1935 and continued over the next three decades with Grand Coulee Dam, the interstate highway system, and the moon landing. There was nothing Americans did not think they could build and virtually no problem for which they did not believe they could devise a technical solution. As the introduction to *Los Angeles: Preface to a Master Plan* noted, "The machine age is here to stay. It must be whipped into condition to serve human needs . . . by planning on a large scale, with technology as a basis."[18] Big engineering projects combined both planning and technology. Additionally, they promised jobs and federal funds for struggling local economies during the Great Depression. Flood control and other engineering works, then, constituted both a material and symbolic boost to faith in national greatness and progress, right at the moment of American capitalism's deepest crisis. Consequently, while the historical conditions of the 1930s militated against nonstructural methods of taming the rivers, those years did bring New Deal engineering to Los Angeles flood control.

With the New Deal, the same depression that had sapped state and local funds for the Flood Control District's Comprehensive Plan in the early 1930s now provided a windfall of federal money. After a false start in 1933, the district finally tapped into this federal largesse in 1935. Under an emergency relief appropriation in that year, the Works Progress Administration authorized the Los Angeles District of the Army Corps of Engineers (LAD) to spend fourteen million dollars to take responsibility for completing fourteen of the Flood Control District's most urgently needed projects. As a result, the LAD staff swelled from fifteen in August 1935 to more than sixteen thousand a year later. The army's involvement rejuvenated the flood-control efforts in Los Angeles, as the workers, 95 percent of whom came from relief rolls, spent a feverish year reinforcing channels, building dams, and digging debris basins above the La Cañada Valley.[19] "No other large WPA project, in California," contended the Army Corps's James Jobes, "has been carried out with more speed, [or] less unit costs . . . than the work done by the [Army] Engineer Department in Los Angeles County."[20]

Between 1936 and 1941, the federal government made the New Deal emergency relief efforts permanent. Federal involvement in flood control nationwide had slowly expanded since the early nineteenth century. In the 1824 case *Gibbons v. Ogden,* the U.S. Supreme Court established that federal responsibility for interstate commerce included river navigation, and Congress subsequently assigned numerous navigational improvement projects to the Army Corps, including limited flood-control works. It was under this navigational authority that the corps diverted the lower Los Angeles River from the harbor in the late 1910s. And in response to devastating floods on the Sacramento, Ohio, and Mississippi rivers in the early twentieth century, and after the 1927 Mississippi flood, the Army Corps intensified its flood-control activities on those rivers. But there was still no national flood-control program. By the mid-1930s, however, there were many additional justifications for such a program, including national defense concerns, the desire to employ jobless workers on public works projects, and New Dealers' aspirations to use public works to stimulate local economic development. As a final impetus, floods devastated many states in 1935 and 1936, inspiring Congress to pass the Flood Control Act of 1936, the first such act of national scope.[21]

Largely because the Flood Control District already had a comprehensive plan ready for implementation, Los Angeles flood control was the first and largest program in the nation to receive funding under the new law.[22] The act included a seventy-million-dollar appropriation for flood control on the Los Angeles and San Gabriel rivers and on the Rio Hondo, a distributary of the San Gabriel River. A 1938 amendment added the Ballona Creek watershed in western Los Angeles and named the entire project the Los Angeles County Drainage Area (LACDA). The acts of 1936 and 1938 directed the Army Corps to survey the watersheds and develop and implement necessary plans for flood control.[23] The acts called for the Flood Control District to cooperate with the army by purchasing rights-of-way, protecting the corps from liability, and maintaining and operating the structures after their completion. Congress authorized construction of the Army Corps's LACDA plan in 1941 and appropriated money for individual projects on an annual basis over the next three decades.

Despite the enticement of federal funding for the local problem, federal engineering did engender some opposition. Supervisor John Anson Ford, for example, considered the Army Corps technically inept and feared cement and materials companies, who stood to profit immensely from river paving, were driving the support for the engineering proposals. Most critically, from his perspective, the funding and attention given to the corps threatened to detract from forest conservation and fire prevention, which he considered cheaper and more important flood-control strategies. The Municipal League echoed Ford's objections. "Engineers," the league's *Bulletin* lamented, "think of flood control problems only in terms of mechanics and hydraulics. The biologic factors and the economic and social aspects are every bit as important."[24] Believing that flood control should be an "integrated part of a regional planning program," the league called for a more varied approach that would include fire protection, water and soil conservation, and zoning.[25] Such local dissension, flood-control supporters feared, might induce Congress, as one official put it, to "spend the money where they aren't so fussy."[26] As Congress debated the Flood Control Act in 1936, the Board of Supervisors chair Herbert Legg directed the Chamber of Commerce and other organizations to offer their unconditional support for the pending legislation.[27] Meanwhile, a southern California congressman admonished Ford that "local opposition would almost to a certainty kill it."[28] Although Ford persisted in raising mild objections to the army's plans, almost everyone else fell into line. After much internal bickering, the Chamber of Commerce supported the plan "in every detail," as did a majority of the Board of Supervisors, the region's congressional delegation, and, of course, the Flood Control District officials.[29] Even the Municipal League muted its objections, admitting that despite certain flaws in the proposal, "these army engineers did do a good job."[30] With such pressure for local unity hushing objections to the corps's involvement, the forestry measures that Ford and the Municipal League preferred got only scant hearing. The various congressional flood-control acts did authorize the U.S. Department of Agriculture to undertake forest conservation measures—fire suppression, erosion control, and tree planting—but over the next twenty years, federal expenditures on forestry work totaled only $10 million, while the engineering projects cost $180 million.[31]

As inevitable as that federal engineering involvement may seem today, it triumphed because the historical conditions of the 1930s rendered nonstructural strategies nearly invisible while making engineering irresistible. The structure of decision-making authority, the growth mentality of southern California, the political imperatives of seeking federal aid, and the pervasiveness of technocratic discourse in public policy circles combined to render nontechnical flood-control approaches virtually unimaginable and prevent them from entering serious debate. In comparison, the path toward more engineering beckoned with promises of federal money, employment for workers, technical expertise of the Army Corps of Engineers, and great public works monuments testifying to American know-how. And all of this would come without requiring hard decisions about limitations on property use, without threatening established political authorities, and without fanning local disagreements over who should pay for flood control and how and where it should be undertaken. In combination, these historical conditions ensured that at the end of the crossroads decade of the 1930s, during which southern Californians confronted flood-control engineering failures and reconsidered how to control the waters of the Los Angeles Basin, they found themselves right where they had started—handing over flood control to engineers to be dealt with solely as an engineering problem. Not only did federal engineering appear compelling at the time; once underway, it gained a momentum of its own, and for the next thirty years—again, for historical reasons—it appeared to be marvelously successful.

THE STORM BEFORE THE CALM

Any doubts about the need for federal engineering drowned as the waters of the great flood of March 1938 exposed the failures of the previous twenty-three years of local flood-control efforts. Approximately one-third of the district's aging temporary works failed in the deluge.[32] And in many places the rivers had little trouble escaping the temporary and undersized flood-control channels.[33] In others, where rainwater ran through the streets and off rooftops in recently developed suburbs, the levees blocked much of the runoff even from entering the channels and instead directed it into neighborhoods and business districts. Meanwhile the dams' drains clogged, and reservoirs silted faster than expected. With the exception of the not-quite-complete San Gabriel Dam No. 1, which heroically captured much of the water from San Gabriel Canyon, few of the mountain dams built with the 1924 bond money appreciably reduced the flood peaks, with outflow from the reservoirs nearly equaling inflow. Despite San Gabriel Dam No. 1's containment of floodwaters, the deluge took forty-nine lives and damaged $40 million worth of property. For both the Los Angeles and San Gabriel rivers, the inundation of March 1938 was the flood of greatest volume on record.[34] As the *Los Angeles Times* lamented, the storm "unerringly picked out the weak spots in our defense and made the most of them."[35] When the famed southern California sun returned a few days after the storm, however, the ever-optimistic Los Angeles mayor Frank Shaw took to the radio. "The sun is shining over southern California today," he reassured listeners, "and . . . Los Angeles is still smiling."[36]

For the next thirty years, the mayor's optimism proved well-founded. During that time, the Army Corps of Engineers reversed the pattern of political stalemate, false starts, and sudden policy shifts that had left Los Angeles so defenseless in 1934 and 1938. The chaos of the local program was replaced by the almost surgical precision with which the army designed structures, set construction dates, secured funding, and built everything on schedule. With 14 contractors, 31 contracts, 2 million cubic yards of concrete, 3.5 million barrels of cement, 20 million cubic yards of earth, 147 million pounds of reinforcing steel, and 920 billion pounds of grouted slope stone, the corps methodically transformed the hydrology of Los Angeles and accomplished precisely what the local district had been unable to do: turn blueprints into structures.[37]

This success earned them widespread praise. In 1965, the American Society of Civil Engineers honored the district with its Award of Merit for outstanding civil engineering achievement, and many agreed with a historian who declared that the dry concrete rivers testified to the "perspicacity of their builders."[38] They were "drainage doctors," one author for *Constructor* magazine said, "hydraulic experts" who brought "huge machines, . . . the finest brains, and modern methods . . . into the fight."[39] Although observers were quick to point out that political bickering and flood disasters had stalled the local program and to praise the expert professionalism of the Army Corps for rising above these challenges, those observers were less apt to recognize the degree to which the Army Corps owed its success to something more than sheer technical proficiency. In addition to machines, brains, and modern methods, the drainage doctors were aided by a prolonged period of political and climatic calm.

The corps's most noteworthy accomplishment was paving the rivers. Although local engineers had contemplated this before, the prohibitive expense of paving the hundreds of miles of channels by hand forced the district in the 1920s to install temporary reinforcements and to implement other cheaper alternatives. Designed to last only a few years, these fortifications were economical in the short run but not in the long. Starting in the 1930s, however, the army put its gangs of relief-roll laborers to work surfacing the riverbeds with concrete. The work accelerated in the 1940s, when technological innovation and the waning of the political will to employ relief-roll workers encouraged the mechanization of

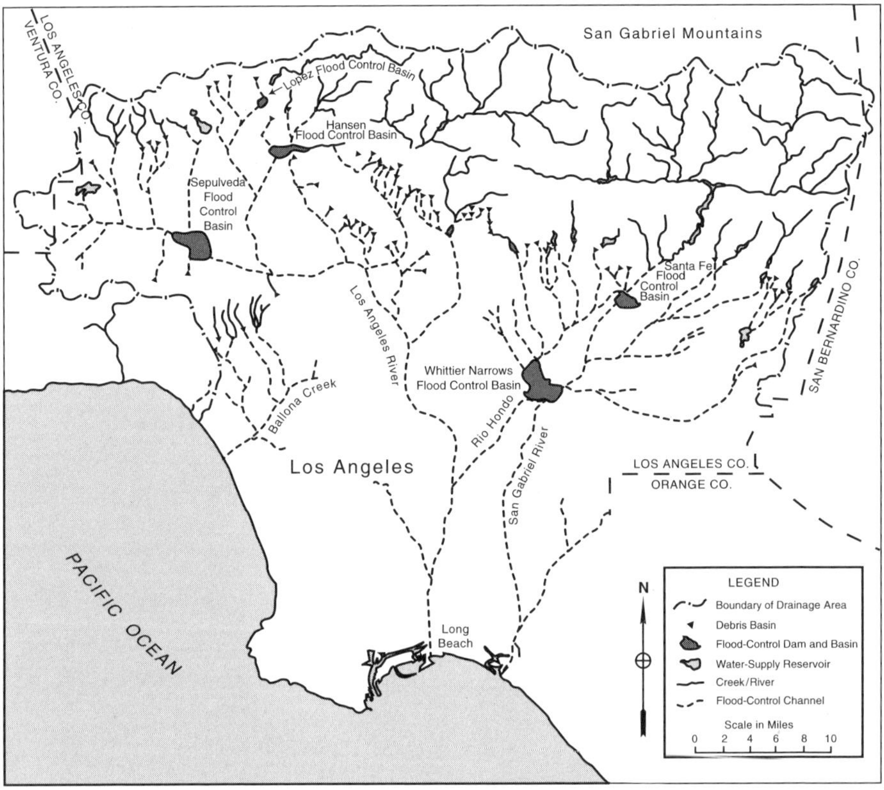

Map 6. Los Angeles County Drainage Area flood-control projects, circa 1970. (Map by Chris Phelps.)

river channel paving. For the next three decades, huge snorting machines—initially on rails and later on giant rubber caterpillar treads—crept through the mostly dry beds of the Los Angeles and San Gabriel rivers, leaving behind them a trail of cement. Whereas workers before World War II could pave two hundred linear feet on a good day, the mechanical pavers deployed at the end of the 1950s could cover three thousand. The Army Corps and its contractors also improved methods of strengthening the channel walls and curving them occasionally to route the channels around obstacles or unacquirable rights-of-way. As of 1937, the Flood Control District had lined seventy miles of channels with concrete or other permanent protection. In the next twenty-five years, however, the faster, cheaper, mechanized methods of paving, coupled with improved knowledge about flow in constructed channels, encased more than three hundred miles of waterways in concrete.[40]

The army also built empty reservoirs on the valley floors. Originally conceived as part of the Flood Control District's Comprehensive Plan, these odd structures, known as flood-control basins, were built to compensate for the inadequacies of the dam and channel program of the 1920s. As the value of riparian property increased, acquiring the rights-of-way to widen channels became prohibitively expensive, and finding some other way of regulating runoff so that downstream channels would not have to be so wide became imperative. Mountain dams, as the 1938 flood demonstrated, were inadequate because there were too few good sites and because they silted up too quickly to provide adequate storage. Moreover, with urbanization more runoff originated on the plain, below the mountain reservoirs. To meet these challenges, the Army Corps built five flood-control basins. Hansen, Sepulveda, and Lopez basins lay on the Los Angeles River and its tributaries, while Santa Fe and Whittier Narrows basins barricaded the San Gabriel. When skies were clear, the rivers trickled through openings in the low, wide dams. During storms, dam operators closed the outlets, and the structures turned into giant funnels that captured the torrents and released only as much as the downstream channels could carry. When the storms ended, operators emptied the reservoirs as quickly as possible to ready the basins to catch another flood if necessary.[41]

With the federal government taking care of the most pressing channel and dam problems, the Flood Control District was able to turn its attention to other matters. Between the 1940s and 1960s, the district constructed extensive water-conservation spreading grounds to supplement the imported supplies on which Los Angeles increasingly depended. It also launched a debris-reduction program for the foothills and installed a variety of works that the Army Corps could not justify as worthy of federal work. Most important, the district addressed the urban runoff problem by taking over the building of storm drains, a responsibility that had previously fallen to individual municipalities. In 1952, county voters approved Flood Control District bonds for the first time since 1924, authorizing a $179 million expenditure to launch a massive storm-drain program. After citizens approved additional bond measures in 1958, 1964, and 1970, the nine-hundred-million-dollar storm-drain program eventually cost more than the main-channel work and produced more than two thousand miles of hidden but vital underground urban plumbing to convey runoff from city streets into the Army Corps's concrete conduits.[42]

In November 1965, when all but one of the projects originally proposed in the Comprehensive Plan were complete, a series of storms showered almost ten inches of rain on southern California in the space of a week. This time, however, muck from the canyons settled into the debris basins fronting the mountains. Clear water streamed out the downhill sides and into paved channels. Runoff from streets and rooftops ran into gutters and then underground drains before joining the flow in the main channels. The waters rose in the paved beds but did not spill over the levees, and flood-control basin operators gauged the weather forecasts and flow data to determine how much to release from behind the barriers. At the end of a week of rain, LACDA's flood-control defenses had spared the region millions of dollars in flood damage and elicited praise from all observers. As the *Los Angeles Herald-Examiner* announced, the structures had "prevented disaster."[43] There were few disruptions to commerce and communication, no walls of rumbling mud, no breached levees, no heroic rescues, no lakes of floodwaters stretching across the plain. For once, water flowed where it was supposed to.

What was most remarkable, however, was that the Army Corps and the Flood Control District accomplished all this with the very blueprint the county had failed to implement in the 1930s.[44] Although the corps tweaked the specifications of various projects, virtually every LACDA project approved by Congress in 1941 had originated as a proposal in the county's Comprehensive Plan. The corps did employ more expensive materials and higher construction standards, and it was also able to take advantage of improved machinery and hydrologic and meteorologic records that spanned longer time periods. But these technical advances cannot by themselves explain the army's success. After all, the difficulties of 1914–1938 had hardly been due exclusively to bad technical designs, and in any case, the army was working with a basic blueprint similar to the one that the locals had. What was different was the set of historical conditions under which the corps labored. In the midcentury decades, in contrast to the 1920s and early 1930s, no graft scandals, rancorous bond elections, flamboyant personalities, bitter meetings of the Board of Supervisors, lawsuits, intraregional rivalries, or newspaper broadsides derailed engineers' plans. There were no sudden shifts in policy, no small problems that exploded through chain reactions into large failures, and no entanglements with other political and ecological systems such as water conservation. There were not even, after 1938 at least, any big floods to disrupt the engineers' plans or sway public opinion against the program. Instead, the political terrain shifted from a landscape of contestation and factionalism to one of consensus in which engineer-bureaucrats held power over an orderly process by which decisions were made and implemented.[45] Certainly the army brought plenty of brains and machines to bear on the problem, but just as the local failures of 1914–1938 stemmed in part from political and climatic troubles, so too did the federal successes of 1938–1969. It was the tranquility of the urban ecosystem, as much as the technical wizardry of engineers, that calmed the waters of the Los Angeles Basin.

One of the most important sources of consensus involved the federal government picking up the check. Although long accustomed to struggling over local flood-control spending down to the last nickel, southern Californians were reluctant to quibble with federal engineering assistance. As the Chamber of Commerce president William Rosecrans noted in 1938, a million dollars "is a lot of money if yours, but if it is Government money it is not very much."[46] But those millions were anything but guaranteed. Many eastern congressional representatives opposed California flood-control appropriations, and the possibility of losing federal funding created intense pressure for local consensus.[47] "Right now the nation is flood-minded," a Los Angeles congressman wrote to Supervisor Ford, asking him to drop his objections to the 1936 Flood Control Act. "In another two or three months, it may become dust-minded, and money for flood control . . . may be impossible to get." He feared that "to buck this tide . . . now is unwise."[48] The congressman and other county leaders recognized that, as Supervisor Herbert Legg put it, Los Angeles would only "get as much as we are united as a county."[49] Consequently, they went to great lengths to maintain that unity. In addition to trying to coax John Anson Ford and other dissenters into silence, county officials created a new public relations division at the Flood Control District and hired professional lobbyists to represent their interests to the U.S. Congress. The county supervisors themselves also made frequent trips to Washington to plead their cause. For these efforts, the Army Corps of Engineers rewarded the Los Angeles County Flood Control District by treating it as the voice of the people. As for the few dissenters, both agencies represented them as voices of minority special interests, whose goals deviated from the good of the many. Any effort to oppose flood control in the era of consensus, then, had to overcome a coalition that included the Flood Control District, the county government, the Army Corps of Engineers, and the usual bevy of private powerful groups and individuals who supported flood control. Against this unified official alliance, potential opponents had little chance of getting Congress to hear their voices. Consequently, almost no substantial challenges to federal flood control arose.[50] And Congress, almost without exception, approved whatever the two agencies put before it between 1936 and 1968. In contrast to the rancorous 1920s, one observer noted in the late 1950s, "local political warfare . . . has in the past twenty years more or less disappeared from the scene."[51]

As important as it was, however, federal money alone cannot explain the muted political conflict of the midcentury decades. As environmentalist challenges to flood-control projects in subsequent decades would later demonstrate, locals could indeed muster opposition even to projects they did not have to pay for. The calming effect of federal involvement on flood-control politics also stemmed in part from the historical events with which this involvement coincided.

Most important among these were the national emergencies that struck between the 1930s and 1950s. The increased federal involvement in Los Angeles was first justified during the Great Depression as a public works project that would employ armies of relief-roll workers.[52] Later, as industrial endeavors that would shortly serve the war effort grew in Los Angeles, local and federal officials increasingly worried about what would happen if a flood were to inundate aircraft factories, oil fields, and steel plants and wash out the lines of transportation and communication that linked them to national defense efforts.[53] In approving LACDA, an author for the engineering magazine *Constructor* observed, Congress was contracting for "insurance that the vital defense industries around the city would not lose time on account of wet feet."[54] To protect that insurance policy, armed soldiers guarded the flood-control sites during World War II. (No saboteurs showed up to menace the defenses, though hundreds of thousands of visitors flocked annually to swim at the reservoirs and gawk at the imposing structures.)[55] Even after the war, however, national defense concerns persisted, as Cold War fears pervaded society and were often invoked as justification for large-scale public works projects, including the Los Angeles flood-control system. Writers of letters to the Army Corps, for example, supported the proposed Whittier Narrows Dam on the grounds that it would protect the region's water supply in the event of "a war, earthquake, or other catastrophe" and that "jeopardy to Coastal Plain Naval installations could affect the national defense of this country."[56] At the same time, national postwar economic prosperity provided enough government money so that flood control and other public programs did not have to compete as fiercely with one another. Thus, a fortuitous combination of historical conditions helped to shield the federal technocracy from the kind of political rancor that had derailed the local flood-control assembly line.

A third important source of consensus was the separation of flood control and water conservation. From the unlikely passage of the 1924 bonds that abruptly shifted the course of flood-control policy to the prolonged disputes over San Gabriel Dam, whenever southern California flood control had intersected with water conservation, something strange had happened. By the 1930s, however, local water conservation was declining in importance.[57] During the 1920s, the federal government began construction on Hoover Dam, which barricaded the Colorado River, stored water for farmers and cities, and generated hydroelectric power for the Southwest, especially southern California. In 1928, Los Angeles, Pasadena, and other southern California cities jointly formed the Metropolitan Water District to share the costs and benefits of Colorado River water, the first of which arrived via a 242-mile aqueduct in June 1941.[58] This outside source diminished the intraregional struggle for local water rights that had ignited lawsuits and federal hearings regarding San Gabriel Dam and made conservation a secondary feature of flood control. Because conservation could not be justified on navigational or national defense grounds, it did not fall within the jurisdiction of the Army Corps of Engineers.[59] Consequently, unlike the star-crossed San Gabriel Dam, in which planners had envisioned both storing water and catching deluges, the flood-control basins the corps built in the 1940s and 1950s remained nearly empty. Meanwhile some of the water that flowed through them was diverted to adjacent spreading grounds, where the Flood Control District let it percolate into underground storage—accomplishing water conservation quite literally on the side. Thus by the 1940s, Los Angeles flood control had for the most part disentangled itself, both physically and administratively, from water-conservation politics.

At the same time that federal funding, national crises, and the separation of flood control and water conservation were eliminating many of the old political controversies, the midcentury reengineering of the rivers was also favored with climatic quiescence. The years 1944 to 1964 brought drought, with all but two years failing to reach the mean precipitation and only a handful of storms of any consequence breaking the dry spell. Record rainfall in 1943 fell on dry ground and as a result registered only a fraction of the runoff that the 1938 storm had produced. Opposite conditions, but similar effects, prevailed in 1952, when light rain fell on saturated ground, causing localized flooding in the San Gabriel Valley and coastal plain. In 1954, rain washed large quantities of debris down from the fire-denuded slopes above the San Gabriel Valley. And in 1956, heavy valley precipitation combined with light mountain rains to produce yet another minor flood.[60] Of course it would have been hard to convince the Arcadia residents who had to drive miles out of their way to cross the inundated El Monte Avenue in 1952 or the foothill dwellers whose houses were smothered by debris in 1954 that these were minor storms. Although these storms were locally intense, however, the flood-control system as a whole actually faired quite well. No flood between 1938 and 1969 matched the fury of those of 1934 or 1938, and the combined damage from the three 1950s storms did not even approach the devastation of 1938. Even had there been no flood-control works, the combined damage costs likely would not have exceeded those of the 1938 deluge.[61] In fact, as an *Engineering News-Record* editor put it in 1956, Los Angeles was extremely lucky to have escaped the last fifteen years of explosive growth and only partial completion of the flood-control works without a major storm.[62] Thus, while the drainage doctors labored to construct LACDA, there was no 1914 deluge or New Year's Eve debris flow to reveal any flaws in what they were building.

Instead, climatic circumstances tended to confirm the value of what was being done and fuel the desire for more of the same. Even the 1938 storm exemplified this pattern. For one thing, it provided engineers with vital hydraulic data. For the first time, an Army Corps engineer noted, "it was possible to observe the action of various flood control structures under the extreme condition of a major flood." With nine governmental bodies gathering data from 730 precipitation stations and hundreds of stream gauges, hydrologists gained their clearest picture yet of the volume of runoff that an areawide southern California storm could generate. From these data they reestimated the severity of the hypothetical fifty-year storm and toughened their design criteria accordingly. Other midcentury floods also taught valuable lessons without substantially altering flood control. Data from the 1943 storm led engineers to refine their design criteria further. The continuing problem of getting the urban runoff into the main channels during the 1950s storms inspired voters to approve storm drain bonds in 1952 and 1958. And the growing cost of cleaning out the debris basins after episodes like the 1954 foothill floods led the Flood Control District to formulate cheaper debris-removal methods. The city of Los Angeles even passed pioneering grading codes in 1952, 1958, and 1963 to alleviate the debris problem.[63] Although engineers learned much from these minor floods, that information was primarily technical data that did not substantially change what they were doing but rather just allowed them to do it better.

Further boosting the confidence of the flood controllers, the damage that did occur was confined to locations not yet served by flood-control installations. As the Flood Control District reported after the 1952 deluge, "In no known case was any damage caused by water after it had entered a permanent flood control channel."[64] Moreover, officials never hesitated to point out, completion of other works on the drawing board would head off even more destruction.[65] "That the flood control system needs to be enlarged is no fault of the engineers," observed the *Los Angeles Times* after the 1938 storm. "Give them the money and they will furnish the additional needed flood protection."[66] All this, coupled with the many successes to which the engineers could point, spawned a new type of flood statistic: damage prevented. After each flood, officials with puffed chests pointed to how much damage the existing structures had prevented, twenty million dollars in 1952, fifty-five million in 1956.[67] The only thing wrong with the flood-control system, it appeared, was that it was not yet done.

Midcentury flood controllers, then, benefited not only from political quiescence but also from a prolonged period of calm waters. The 1914 flood, which caught southern Californians off guard, spawned the first organized local flood-control program, and the New Year's Eve debris flows of 1933–1934 heralded the downfall of that regime. Even the relatively minor geologic event of the San Gabriel Dam landslide in 1929 had revealed the infeasibility of the district's central project and led to the recasting of the world's greatest dam as a smaller two-dam project. In contrast to these events, which significantly altered flood-control efforts, none of the floods in the midcentury decades challenged the political or technical framework of southern California flood control. Instead, the cumulative effect of the small floods was to encourage continuation of existing practices, affirm the apparent efficacy of those efforts, and resuscitate voter generosity in bond elections. In tandem, the political and climatic quiescence between the 1930s and 1960s provided the drainage doctors with favorable conditions under which to execute their designs. It is little wonder, then, that the Army Corps and the Flood Control District accomplished everything they planned between the 1930s and 1960s: there was nothing in nature or society to stop them.

WHITTIER NARROWS: THE EXCEPTION THAT PROVED THE RULE

If political quiescence and technical accomplishments were the rule in the midcentury decades, then the conflict over Whittier Narrows Dam was the exception. First proposed in the 1931 Comprehensive Plan, the dam was included in the 1933 petition for federal aid, withdrawn from federal consideration in 1935 because of public protest, and restored to LACDA in 1938. After World War II ended, a bitter struggle over the project delayed its completion for a decade.[68] In comparison to the consensus that characterized virtually every other LACDA project, the controversy over the Narrows appears to be a significant exception. Yet it also proved many of the rules that defined midcentury flood control. First, as both the only case in which substantial local disagreement arose and the only case in which the drainage doctors experienced difficulty implementing their designs, the Narrows controversy underscores how important political quiescence was to the engineers' construction progress elsewhere in the county. Second, the fact that the conflict, unlike the San Gabriel Dam fiasco, did not spill over into other projects and disrupt the entire flood-control program illustrates the stabilizing effect that federal participation had on flood control. Finally, the Whittier Narrows controversy exemplifies the midcentury continuation and expansion of technocracy as the dominant policy-making paradigm for southern California flood control.

Whittier Narrows Dam was the first of the flood-control basins to be proposed. At the southern end of the San Gabriel Valley, the San Gabriel River crossed a low range of hills through a gap known as Whittier Narrows. There, the land had folded, faulted, tilted, and eroded, and an impermeable underground rock formation forced the groundwater of the San Gabriel Valley to the surface.[69] The Rio Hondo, an old channel of the San Gabriel River that branched off several miles north, also flowed into the gap, nearly rejoining the San Gabriel as the two ran along opposite sides of the Narrows. Through this gap flowed all of the runoff and groundwater of the San Gabriel Valley.

As the gateway to the coastal plain, Whittier Narrows posed both a water-conservation opportunity and a flood-control threat. Because much of the water passing through the Narrows originated as valley runoff below the mountain dams that the Flood Control District was building in the 1920s and 1930s, much of this water was uncontrolled, threatening the downstream coastal plain with inundation. Moreover, merely channeling the waters passing through the Narrows without letting them percolate into the permeable soils on the coastal plain constituted a missed opportunity for water conservation. A dam at the Narrows, then, not only promised to regulate a substantial portion of the runoff entering the coastal plain but also offered conservation benefits as well. The district proposed such a dam in its Comprehensive Plan of 1931. Except for a small pool of water, the reservoir behind the dam would remain empty, ready to catch a flood and release it slowly. The size of the channels below the outlet could be made considerably smaller because of the protection the dam would afford. The slowed, desilted waters could be percolated into the downstream riverbeds more effectively than without the dam, and excess waters would be diverted into adjacent spreading grounds.[70] When the army included Whittier Narrows Dam in its 1938 survey report and Congress approved the project as part of LACDA in 1941, little opposition arose. The project was federally funded. It separated flood control and conservation. And it promised to protect several aircraft factories and a U.S. naval air station.[71] It seemed to epitomize the new consensual political landscape.

El Monte residents, however, objected. A town of about fifty thousand, El Monte lay in the spot that the district and the Army Corps proposed to inundate occasionally behind the Whittier Narrows Flood Control Basin. Situated upstream from the Narrows and protected by the San Gabriel dams and the soon-to-be-completed Santa Fe Flood Control Basin, El Monte had little to gain from the Whittier Narrows project and much to lose. Local business stood to lose clients. Three thousand residents stood to lose their houses. School districts feared losing investment in recently improved properties in the proposed inundation zone.[72] The Reverend Dan Cleveland, a wealthy Democratic Party supporter, pastored a church that stood to lose some of its membership. And the area's Democratic congressman, Jerry Voorhis, who sat on the House Flood Control Committee, feared losing support from his constituents if Congress were to approve the project. Spurred by these potential losses, the town's Chamber of Commerce formed the Anti–Whittier Narrows Dam Association, a coalition of government officials, businesses, and powerful private individuals such as its chair, Rev. Dan Cleveland. With such politically articulate leadership, El Monteans were able to crack the technocratic alliance of engineers, government agencies, and development interests that favored the dam, and Whittier Narrows became the most notable exception to the rule of consensus in the era of federal flood control.

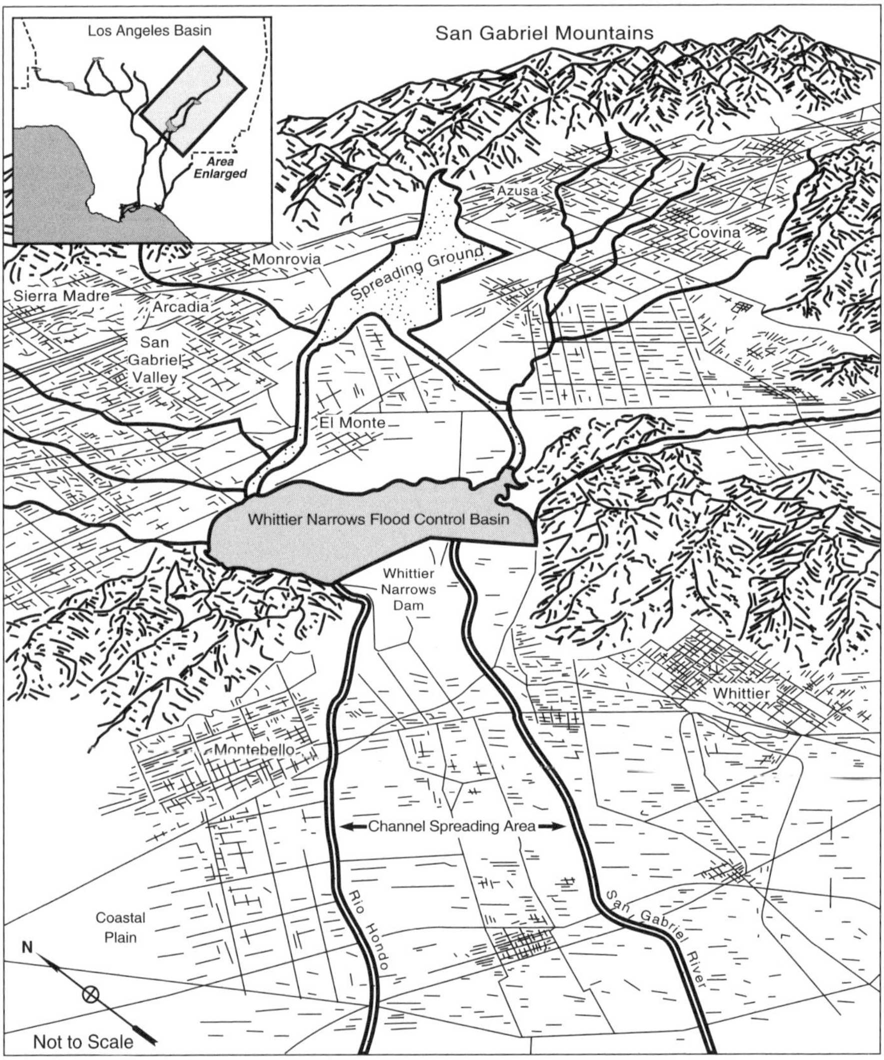

Map 7. Whittier Narrows and vicinity, showing the dam and flood-control basin as first conceived by the Los Angeles County Flood Control District in the early 1930s. (Map by Chris Phelps.)

The project also, therefore, became the single item in LACDA on which progress stalled. In response to the pressure El Monte levied against the Board of Supervisors, the Flood Control District removed the dam from its Comprehensive Plan in the 1935 application for federal funding.[73] The downstream landowners, chambers of commerce, and municipalities that favored the flood protection and the cheap water conservation the dam would provide then persuaded the district and the army to restore the Whittier Narrows Flood Control Basin to consideration. In the late 1930s and early 1940s, the question became largely moot, because the corps could not build the project with the appropriations of the 1936 and 1941 Flood Control Acts and because during World War II the army halted all nonmilitary engineering projects, including Whittier Narrows Dam.[74]

In the meantime El Monteans armed themselves for technical battle. They hired a team of engineers including the retired James Reagan to devise a damless alternative that came to be known as the all-channel plan because it proposed to widen the San Gabriel River and Rio Hondo in lieu of a dam at the Narrows.[75] Calling the corps's proposal reckless and criminal, El Monteans promised that their all-channel plan would control nearly two hundred thousand cubic feet per second more than Whittier Narrows Dam and that it would do so for less money and without taking any property.[76] El Monte's engineers further claimed that as locals they had a more expert familiarity with the problem than the army's engineers, and they disputed the army's estimates of the maximum possible rainfall, the duration of peak flow, and the cost of reconstructing the channels. They attacked the army for its designs that had failed in the 1938 storm. And they accused the army of basing its stance on "mere assertions" and "fundamental error."[77] The El Monteans and their engineers did not disagree with the technocrats' assumption that the region's rapid urbanization necessitated flood-control structures to protect all the development; they merely objected to the specific methods.

The proponents' response was equally technocratic. In December 1945, after the war ended and the Whittier Narrows controversy resumed, the Flood Control District chief Harold Hedger compared the costs and benefits of the army's flood-control basin plan and the El Monteans' all-channel proposal. In addition to providing inadequate protection, in Hedger's opinion, the forty-six-million-dollar all-channel proposal would double the cost of the army's plan. Asking Congress to spend the extra twenty-two million dollars, Hedger feared, would result in the "failure to obtain Congressional appropriations for either system, which would indeed be disastrous."[78] The debate was beginning to sound like that of San Gabriel Dam all over again: two deadlocked coalitions of governments and privates interests, with both sides claiming to represent the public good and both sides marshaling experts armed with technical data to support their cases. Indeed, many people such as Hedger feared that the Whittier Narrows controversy had the potential to stifle the whole flood-control program, just as the San Gabriel Dam fiasco had.[79]

In contrast to the local program that the San Gabriel conflict had overwhelmed, however, the federal flood-control assembly line was more tolerant of unforeseen changes. An Army Corps flood-control project began when local interests asked their congressional representative to include it in a flood-control bill. Congress then requested a report from the War Department, and the secretary of war referred the project to the Army Corps's district engineer in the region where the proposed project was to be located. Next the district engineer prepared a preliminary report on the engineering and economic feasibility of the project and forwarded the report up through the chain of command to the corps's chief engineer in Washington, to the Board of Engineers for Rivers and Harbors, and finally back to the secretary of war. A proposal that made it this far then went to the House Flood Control Committee for consideration. After congressional and presidential approval, the plan went back to the local Army Corps office for a complete survey report and the formulation of a definite project plan, which again had to ascend the army hierarchy. In all, there were thirty-two separate steps at which opponents could try to block a given project. Even at the end of the process, the possibility remained for changing or challenging the project, because Congress authorized funds for projects only every few years, meaning that even after a project was approved, opponents might later have the opportunity to block appropriations.[80] In the case of Whittier Narrows, for example, Congress approved plans in the Flood Control Acts of 1938 and 1941 but did not appropriate money for it on either occasion, opening the possibility for El Monteans to oppose appropriations after the war.

This procedure departed sharply from California's Flood Control Act of 1915. That law had required only a preliminary report by the Los Angeles County Flood Control District and a vote by the citizens to approve a project. Once a plan had been approved by the voters, it was illegal even to study alternatives without a court order acknowledging a change of conditions. Because opponents had no legal mechanism for modifying projects, they had resorted to attacking the entire flood-control program through newspaper broadsides, campaigns to overthrow the chief engineer, opposition to bond elections, competing federal right-of-way petitions, and lawsuits against the Flood Control District. Although all of these types of attacks had stemmed specifically from San Gabriel Dam, the effects of the conflict generated chaos throughout the district's program and so thoroughly discredited it that it could not get funding for its other projects. Thus controversy over a single project had stalled the entire local program. The federal process, in contrast, offered anyone with the right political connections and the ability to put forth highly technical arguments a chance to approve, disapprove, or modify plans and to exert pressure at every stop on the assembly line. Arcane as it was, this procedure contributed to the midcentury political quiescence by containing individual project controversies within bureaucratic hierarchies and thus preventing them from spilling over and affecting the rest of the program.

Nowhere were the benefits of this policy-making procedure more evident than in the case of Whittier Narrows. The storm of 1943 provided new rainfall data that altered the corps's estimation of the maximum possible rainfall and led the engineers to toughen their design criteria. Essentially this was a changed condition, analogous to the landslide that had revealed the unsuitable geology for the San Gabriel Dam foundation. On that occasion, it had taken dozens of engineering reports, an order from the state engineer's office, and a court approval before the district could proceed with alternative plans. The army, in contrast, had little difficulty responding to the technical challenges the storm posed. In 1944, the corps's Los Angeles officers completed a new preliminary report and forwarded it up the chain of command with a request for permission to revise the plan. In the new version of the project, engineers upgraded the basin's design outflow from twenty-two thousand to forty thousand cubic feet per second and redesigned the dam so that the basin would discharge floods primarily down the Rio Hondo and into the Los Angeles River, which had a larger channel than the lower San Gabriel River, the outlet previously proposed. After receiving permission from his superiors, the district engineer proceeded with the new plan.[81]

When the corps prepared to seek congressional funding for this new plan in 1945, however, El Monteans again resisted. A man who lived in the proposed inundation zone revived the dormant controversy in January 1945 by writing a letter of protest to Representative Voorhis.[82] The carbon copy the man sent to California senator Hiram Johnson found its way to the Army Corps's chief of engineers in Washington and eventually back to the corps's Los Angeles office, at which point the pro-flood-control machine swung into action. The army's Los Angeles office prepared a detailed history and description of the Whittier Narrows controversy and assured the chief in Washington that the protesters did not represent the will of the majority of the people. In keeping with its alliance with the army, the Los Angeles County Flood Control District endorsed the corps's plan. After the Army Corps completed its new definite project report in April 1945, resolutions also poured into the army offices from communities and businesses downstream of the Narrows in support of the dam. Both supporters and opponents continued bombarding the army's engineers and other officials with letters and resolutions as the definite project report worked its way up the chain of command. In October, the El Monteans sent representatives to oppose the dam before a House Appropriations subcommittee. Although the Army Corps, the Flood Control District, the Board of Supervisors, the state engineering and water resources departments, and every downstream municipality supported the Narrows dam, Congress was unwilling to fund a project in the district of a representative who opposed it. Congress appropriated no funds and instead, in 1946, directed the army to restudy the whole issue, including the alternative all-channel plans.[83]

On 12 December 1946, after the corps had investigated the alternatives further, angry El Monteans packed a public hearing held in downtown Los Angeles, at which the army presented its findings. The crowd cheered every point made by El Monte's advocates and booed the dam's supporters.[84] On one hand, as the corps held fast to its dam proposal, the raucous display exemplified the almost insurmountable obstacle the technocratic alliance of government officials, technical experts, and private interests posed to anyone who challenged the flood-control establishment in the midcentury decades. In its report on the hearings, the Army Corps emphasized the unanimous support for its own plans among state and local agencies, "which are responsible by law for the safety and welfare of the people," and among numerous "semi-public groups," including water companies and chambers of commerce. The corps cast the opponents, in contrast, as "a relatively small local unofficial group" and pointedly observed that they represented "an area which has been substantially protected from flood damage by previously constructed flood-control works."[85] Although in reality the composition of the El Monte coalition did not differ substantially from the prodam forces, the corps adopted rhetoric reminiscent of the Progressive Era in order to portray the El Monteans as an insignificant and selfish minority opposing the needs of the many. On the other hand, however, the hearings also revealed why the El Monteans had been able to make themselves an effective exception to the midcentury flood-control consensus. Their ability to advance their arguments in technical terms had enabled them to force the hearing in the first place, and their political connections enabled them to maintain congressional favor even after the hearings, when the Army Corps continued to balk at the all-channel plan. As a result, the midcentury federal flood-control assembly line, so irresistible everywhere else in the region, stalled at Whittier Narrows, only because it met an equally powerful and equally technocratic opposition.

The deadlock finally ended a year later as a political bargain. When the powerful California senator William Knowland demanded a resolution to the conflict in conjunction with a larger water-resources bill he was arranging for the state of California, freshman representative Richard Nixon, who had defeated Voorhis in 1946 and who was now up for reelection, quickly brokered a deal among the various interests. According to the agreement struck in March 1948, the army would build Plan B, its second choice among the six alternatives presented at the 1946 hearings. The dam site would be moved downstream to minimize the potential inundation of El Monte, and levees would ring Temple Elementary School and a Texaco oil and gas refinery to exclude them from the retarding basin. Congress funded the project in 1949; construction began in 1950; and the Whittier Narrows Flood Control Basin was dedicated in 1955. Although more expensive than the army's 1945 plan, the compromise version was considerably cheaper than the all-channel proposals, and both sides were satisfied. Nixon gained credit as the heroic arbitrator, and even the Democrat Rev. Dan Cleveland campaigned for him in the election.[86]

Although of only minor significance to Nixon's career, the Whittier Narrows saga and the eventual compromise illustrate the central distinctions between flood-control politics of the early local period and the federal era. Whereas the San Gabriel Dam conflict became a runaway train that derailed flood control in the 1920s and 1930s, the Whittier Narrows dispute remained within federal procedures for modifying projects. Those procedures offered El Monteans numerous opportunities to oppose the project, and they had the necessary political and technical resources to gain a voice in the technocratic flood-control debate. Significantly, though, the procedures also offered them channels to voice their concerns without resorting to the lawsuits and the anti–Flood Control District election campaigns that had caused the San Gabriel Dam fiasco to fester until it discredited the entire flood-control program. Finally, the decision-making procedure provided for a political compromise, something 1920s flood control did not allow. The San Gabriel

Dam contest had unmasked an inflexible policy-making system in which problems arising early in the procedure multiplied as they moved along the decision-making path until they caused a system failure. The federal operations, in contrast, were able to advance flood control on all other fronts until locals resolved their differences. Consequently, although the Whittier Narrows conflict delayed the project for seventeen years, it did not jeopardize the rest of the flood-control program.

THE NEW PATH FROM SKY TO SEA

If LACDA's dams and channels had been in place in 1938, the Army Corps's colonel John Dillard declared in 1965, the disaster would not have occurred.[87] Indeed, the path of the water from sky to sea had radically changed since the 1930s. Before it could damage the city, a flood striking at the end of the 1960s would have had to fill the 106 mountain debris basins, spill over the five valley flood-control dams, or make a breach somewhere along the 350 miles of concrete river channels.[88] Flood control had come a long way from the rotted wood and rusted automobiles that lined the channels in 1938. Observers celebrated this transformation as a triumph of technical expertise, but in reality, it also resulted from political and climatic quiescence. As the Whittier Narrows struggle shows, wherever locals disagreed, flood-control progress stalled. That delay, however, ultimately proved of little consequence. The flood-control program could afford a seventeen-year hiatus in one of its major projects without suffering serious consequences because of the environmental quiescence. Conceived in 1931, Whittier Narrows Dam was not completed until 1955, but its services were not needed during those years.

CHAPTER 6

Necessary but Not Sufficient

Storms, Environmentalism, and New Visions for Flood Control, 1969–2001

In 1969, LACDA lived up to John Dillard's promise. Between 18 and 26 January, rain fell almost continuously. The dams overflowed; streets and buildings flooded; debris rumbled out of the mountains and buried seven people alive in their beds. Damages totaled thirty million dollars; the death toll, seventy-three. When it was all over, nearly thirteen and a half inches of rain had fallen on downtown Los Angeles, more than in any other nine-day span in southern California's recorded history. By comparison, eleven inches had fallen downtown in 1938, eight in 1934, and seven in 1914. Although some areas suffered considerable destruction, LACDA performed well on the whole. Engineers estimated the prevented damages at more than a billion dollars.[1] "The Los Angeles County drainage area project," the Army Corps declared, "protected the Los Angeles metropolitan area from what otherwise would have been unprecedented damage."[2] As they had done with previous storms, flood controllers assessed the 1969 gale as proof of the structures' success.

Their euphoria had faded, however, by the mid-1980s. In a "LACDA Update," which the Army Corps mass mailed to Los Angeles County residents in September 1987, the corps warned, "Disastrous Flooding Could Return to Los Angeles County."[3] Sediment was building up behind dams faster than expected, decreasing their protective capabilities. The concrete channels had eroded. Flows in rivers had surpassed expectations. So had the pace of urbanization. The flood-control system itself had increased the peak discharge in major channels. A map in the brochure depicted the areas where the army expected inundation during a large deluge. "The existing flood control system is no longer capable of protecting hundreds of thousands of residents from large floods," the bulletin warned. "Some of them are you!"[4]

The brochure also exhibited a new approach to the process of flood-control policy making. Although it avowed that "technically, engineers can design just about anything," the bulletin also acknowledged that "beyond what will technically work, planners must listen and pay attention to resident ideas, suggestions, desires, objections, and concerns." The corps asked readers, "How do you feel about the potential flood control measures?" What "potential solutions do you want them [the engineers] to consider?"[5] In a departure from a half century of treating flood control solely as an engineering problem to be dealt with solely by engineers, the experts now called for public involvement.

This change reflected two decades of political and environmental turmoil. First, in the late 1960s and early 1970s, environmentalists began mounting the first serious political challenge to flood-control engineering since the 1920s. Then in 1978 and 1980, comparatively minor storms produced catastrophic damage, leading engineers to rethink the technical soundness of LACDA. Together, the environmental activism and the renewed flood problems reopened the entire question of flood control for the first time in seven decades and injected nontechnical strategies and public involvement into flood-control planning. Although the old flood-control regime of reengineered rivers and expert bureaucrats by no means collapsed, the political and environmental turmoil after 1969 shook the foundations of the flood-control program that southern Californians had instituted in 1915.

THE POLITICAL CHALLENGE: ENVIRONMENTALISM

The Los Angeles County Flood Control District's *Biennial Report* of 1969–1971 displayed an unmistakable shift. After decades of reports that described weather conditions, technical matters, and accounting balances, that year's edition added a new category to the topics it covered: environmentalism. Never before had a *Biennial Report* mentioned concerns such as cultural values, recreation, aesthetics, or the environment, and yet this volume was filled with such references. "With the growing public interest in protecting and enhancing our environment," the report announced, "the District has implemented a new program of holding public meetings." Through these hearings, the Flood Control District sought not only to inform citizens but also "to make our engineers sensitive to possible social and environmental problems of each project." Also during the biennium, the district began filing federal- and state-mandated environmental impact statements for new projects and initiated a landscaping program for the "aesthetic improvement" of existing ones. It undertook these new practices as part of its overall goal to work "in harmony with the best interests and prevailing values of the people."[6] All of this sounded very *un*technocratic.

The 1969–1971 *Biennial Report* also noted that LACDA was entirely complete except for one project, Sierra Madre Wash, which flowed out of Little Santa Anita Canyon.[7] As the lone exception to the army's remarkable midcentury track record of building whatever it planned, the case of Sierra Madre Wash exemplified environmentalism's challenge to technocracy. In the 1960s, Sierra Madre was a town of twelve thousand nestled in the foothills of the San Gabriels thirteen miles northeast of downtown Los Angeles. Sierra Madreans fancied their town as just a little bit different, and they were right. While freeways, malls, chain stores, and fast-food restaurants had invaded suburban Los Angeles during the 1960s, Sierra Madre boasted an ecology fair, a community organic garden, only one stoplight, and one of the nation's first urban wilderness parks.[8] With its inexpensive rents and picturesque scenery, the town had attracted a thriving community of artists, who had recolonized the shacks and cabins of a turn-of-the-century mountain resort and sanitarium at the mouth of Little Santa Anita Canyon. The babbling cobblestoned streambed of Sierra Madre Wash meandered through the well-wooded rustic canyon neighborhood. A favorite haunt in the late 1960s was the Canyon Store, with its old potbellied stove. There, residents would gather to kibitz about community affairs, including the Army Corps's plan to channelize their beloved Sierra Madre Wash.

In 1927, the Flood Control District had dammed the mouth of Little Santa Anita Canyon and constructed the sinuous one-mile rubble channel that by the 1960s the canyon residents ironically prized as being natural. Even with these structures, however, the capacity of Sierra Madre Wash was inadequate to carry the flow of the corps's design flood, and so the 1941 Flood Control Act contained a plan to improve the channel.[9] Reconstruction was slated for the mid-1960s. Like other tributaries the LAD had redesigned in the 1940s and 1950s, the improved Sierra Madre Wash was to be a reinforced concrete box channel, paved on three sides, open on the top, and designed to carry a larger capacity than the existing channel could. It would be lined with chain-link fence. Rarely in the previous two decades had the corps encountered any environmentalist opposition to such projects, and it had rebuffed those scattered pleas for more aesthetic designs on the grounds that covering or landscaping the flood-control channels was unsafe and uneconomical.[10] There was no reason to suspect that the Sierra Madre Wash would be any different.

The lack of concern for the ecological impact of paving rivers was hardly surprising, for midcentury environmentalists in Sierra Madre and elsewhere in southern California had their minds on other things. The region's Sierra Club chapter, for example, was basically a private outings club. To join this exclusive group, one had to be sponsored by someone already inside, and the McCarthy-era leadership even toyed with the idea of requiring that members take a loyalty oath stating that they did not belong to any subversive organizations. One faction within the chapter also favored excluding African Americans altogether.[11] The main activities of the group consisted of weekend hikes, slide shows, and other get-togethers. The national organization even threatened at one point to expel the chapter for being "too playful."[12] To the extent that the chapter concerned itself with conservation at all, it focused on threats to Dinosaur National Monument, the southeastern California desert, and other faraway wildernesses, and paid no attention whatsoever to local urban environmental questions.[13] For environmentalists, then, just as much as for Flood Control District and Army Corps officials, the rivers of southern California were strictly an engineering problem.

By the time the engineers came to Sierra Madre in 1965, however, things were different. Since the Progressive Era, Americans had valued the objectivity and efficiency that they believed came with expert planning. John F. Kennedy exemplified this faith when he said that America's remaining problems "are technical problems" that involve "sophisticated judgments which do not lend themselves to the great sort of 'passionate movements' which have stirred this country so often in the past."[14] Even as he spoke these words, however, Americans were proving him wrong. Since World War II, Americans had been told they were living in the best of societies. Education, technology, affluence, military might would solve all problems. And yet when Americans looked around in the early 1960s, they saw illiteracy, poverty, pollution, violence, urban decay, racial inequality, disillusioned suburban women, and a nation living in fear of nuclear war. Many people decided to do something about it. Drawing from the Progressive Era settlement house movement and leftist community organizing launched by Saul Alinsky and others in the 1930s, grassroots organizations sprang up everywhere, seeking to organize people around every conceivable topic.[15] Health care, day care, housing, recreation, food co-ops, senior citizens programs, poverty, jobs, schools, pollution, occupational safety, high rents, taxes, racism, sexism, public services, parks, slow growth, crime, nuclear power, police violence, immigrant services, drugs—the list of causes was endless. By the 1970s, the *Christian Science Monitor* noted that the nation was swept up in "a groundswell movement of citizens," and by the 1980s, the historian Robert Fisher later declared, "voluntary citizen action organizations" had become the "dominant strategy to address social problems."[16] By 1990, there were two million citizen action groups in the United States, and the movement had gone international with the proliferation of nongovernmental organizations.[17] Despite the stunning diversity in issues, tactics, and membership among these groups, most shared the belief that people should have a say in the policy making that shapes their lives. According to Fisher, one of the themes growing out of the community-organizing movements was "the demand for autonomy from technocratic control, stemming from the belief that people have the ability and the right to control their day-to-day lives."[18] Or as one activist put it, "We believe in democracy—and average people controlling their lives."[19] Such a commitment required challenging not only the content of many public policies but the process of policy making as well.

One of the grassroots movements that sought to challenge both policy and policy making was environmentalism. During the 1960s, longstanding but diffuse strands of American concern for the environment focused into a political movement. In 1962 Rachel Carson published *Silent Spring,* alerting the public to the chemical killers lurking in the environment. Six years later, the 1969 Santa Barbara Channel oil spill splashed pictures of tarred birds and blackened beaches across the covers of newspapers and magazines nationwide. Along with rising postwar affluence, these dramatic events led middle-class Americans to value the "beauty, health, and permanence" of their lives and surroundings and thrust concerns about the environment, recreation, and aesthetics to the fore of American politics.[20] Membership in the Sierra Club and other environmental organizations skyrocketed. Citizens elected environmentalist public officials, such as Wisconsin's Gaylord Nelson and Arizona's Morris K. Udall, and supported legislation, such as the 1964 Wilderness Act and the 1968 Wild and Scenic Rivers Act. They demonstrated against pollution, occupational hazards, nuclear plants, and hydroelectric projects. By the time all of this activism peaked at the national Earth Day celebration on 22 April 1970, environmentalism enjoyed a national constituency of politically connected people willing and able to voice the movement's convictions.[21]

Many southern Californians energetically joined this national movement. In addition to fighting offshore oil drilling, foothills mining, freeway construction, urban sprawl, and inner-city blight, their concern began to increase for the health of their rivers.[22] In 1961, one writer lamented that the Los Angeles River, which once ran "where it chose, climbing over mossy, jutting boulders, descending in miniature waterfalls glinting with color as sun and water joined," now flowed "in a disciplined line bordered by miles of cemented beds, guarded by flood dikes, caged in wire fencing."[23] Five years later, another author described the rivers as a collection of "underground tunnels, metal tubes, and street gutters" that had been "dynamited, bulldozed, widened, straightened" and whose "scenic beauty is gone."[24] The rivers are doomed, a third article declared. "Gradually, and unnoticed, they have disappeared one by one with hardly a protest from the city dwellers."[25] To prevent further destruction to the rivers, Sierra Madre Sierra Clubbers, along with members of several other civic organizations, mounted a protest against plans to fill Little Santa Anita Canyon with check dams in the early 1960s.[26] Into these more politically charged circumstances, in 1965 engineers innocently introduced the plans to improve Sierra Madre Wash.

One thing the engineers had not counted on was the degree to which environmentalists would insist on participating in flood-control policy making. Like the rest of the nation's grassroots revolution, environmentalism challenged the alliance of technical experts, government authorities, and private interests that had dominated policy making in so many areas of American life since the Progressive Era. By the 1970s, environmentalism had whittled away much of that technocratic authority, forcing the American policy-making system to make room for citizen activism and to include aesthetics, health, and other nontechnical concerns in its calculations.[27] Moreover, state and federal legislation in the 1960s and 1970s erected a legal framework that required public input into a variety of environmental policy areas and provided citizens avenues for challenging policy making that ran afoul of this new openness.[28] In response, the Army Corps and the Flood Control District began to take steps to gather public opinion on their projects. "In recent years," the corps's colonel Ken Roper acknowledged in his opening remarks at a 1972 public hearing, "a major attitude change has taken place in the desires and attitudes of the public. Along with you, we are concerned with what are termed ecological and environmental values." In short, Roper continued, "we want public input."[29] Although Roper and other officials learned to tailor their rhetoric to fit the new political realities of citizen activism, the grassroots revolution required more than rhetoric. The result was a struggle between engineers, who were accustomed to controlling policy making, and environmentalists, who demanded that new ideas and new decision-making processes be incorporated into the flood-control program.

Nowhere was that struggle more contentious than in Sierra Madre. In 1966, opponents of the Sierra Madre Wash plans persuaded their congressional representative to attach to the 1966 flood-control appropriations bill a rider that granted the city the right to review all phases of the project. Meanwhile, the environmentalists brought considerable pressure to bear on the city council. The council, torn between its desire to appease the canyon residents and its duty to protect the community from flooding, followed a tortuous path, on one occasion endorsing the engineers' plans, only to rescind the approval a few weeks later. Several times the frustrated engineers threatened to wash their hands of the matter entirely and leave flood-prone Sierra Madre to its own fate, but each time they came back with another proposal.[30] By March 1967, when the Flood Control District's Omer Hall and the army's Fred Cline met with the council at a public meeting, tensions were high, and each side doubted the good faith of the other. At the meeting, the engineers dodged certain questions and invoked their expert credentials to avoid having to explain their answers to others. When council members probed them about technical questions—what alternatives had they explored? what had they found? why did the two agencies' estimates of the maximum possible flood size differ?—the engineers grew defensive. Hall accused council member Jean Maddox of doubting his expertise. Exasperated by the engineers' dissembling, the council members grew blunter until Hall finally snapped, "The Corps['s] responsibility is to design the project and ours is to operate and maintain that project, and yours is to give us your approval to that project. Now, without those three things, we are not going to have a project."[31] Experts decide, the public submits—that was the way flood control had worked since 1914. When one council member agreed with Hall, however, and suggested that the city officials were not "competent to sit in judgement of the engineering," Maddox retorted, "I'm afraid I can't go along with that. I think that we still have the right to see the facts and figures upon which these agencies have based their conclusions."[32] Whether Hall liked it or not, Maddox and her environmentalist supporters were insisting that flood control cease to be a problem to be dealt with solely by engineers.

Furthermore, in their eyes, it was no longer solely an engineering problem. Since 1914, the assumption that underlay flood-control efforts in Los Angeles had been that reengineering the rivers was necessary to allow the development of Los Angeles. As late as 1971, in its report on new proposals for projects in the Santa Clara River valley in northern Los Angeles County, the Army Corps had stated that "without flood control, substantial areas in the flood plain cannot be developed to their full potential. Optimum and orderly development of the flood plains is required to meet the growing needs of the metropolitan area."[33] Environmentalists testifying at Colonel Roper's 1972 hearing, however, offered competing definitions of both flood control and development. One homeowner advocated the preservation of the "natural character" of the stream that ran near his home. He and other residents, he said, "do not want that creek channelized, even if it does mean some property damage." Others called for landscaping and the construction of parks, bicycle paths, and horseback riding trails along the Flood Control District's rights-of-way. Meanwhile, the Sierra Club's Angeles Chapter conservation chair Cecile Rosenthal labeled such concerns an "expansion of use" of flood-control facilities and challenged the Army Corps's definition of its own mission. "The Corps," she acknowledged, "is charged with determining 'optimum development of all water and related land resources in the Los Angeles County Drainage Area.'" But optimum development, she argued, did not necessarily have to entail expansion of the channels and drains. Instead, she suggested, "optimum development could be not interfering with the natural replacement of beach sand. Optimum development could be keeping natural drainages as wilderness, wildlife habitats, watershed, water source, educational and scientific resources, and for aesthetic purposes." Flood control that allowed subdivision of the land, she continued, did not necessarily profit the community. Development brought new costs for services such as firefighting, schools, police, and roads.[34] The urban development that southern Californians had long considered a benefit, Rosenthal and other environmentalists now counted as a cost.

Sierra Madreans could not have agreed more. As the city council resolved in 1966, the wash was "not just a typical watercourse" but rather "an integral part of development and flavor of Sierra Madre." The wash ran "through a part of [the] city that has a particular rural character," council member Ed Wagner said in 1967. "To put a concrete ditch there would completely destroy it." The conflict dragged on for ten years as residents pressured the district and the corps for an alternative plan that would retain the rustic character of the neighborhood and be more aesthetically pleasing than the reinforced concrete box channel that the engineers planned to build. The two agencies balked. They considered preservation of the neighborhood's character and aesthetics to be mere "embellishments," on which they were willing to spend a small amount of money but not much. "Our obligation," one army engineer informed the city council, "is to design . . . the lowest cost facility that will meet the [flood protection] requirements of the project." The town itself would have to finance any "embellishments" that increased the cost of the project by more than 3 percent.

Environmentalists dug in. They circulated anti-flood-control petitions (one signed by more than twelve hundred residents), elected Maddox and Wagner to the city council, enlisted the aid of the local congressional representative, and attended every city council meeting at which flood control was discussed. Sometime in the early 1970s, one environmentalist remembered, residents who gathered around the potbellied stove at the Canyon Store organized a group called Save Our Stream and sold T-shirts to publicize the battle. The engineers called them crazy and suicidal, but the Sierra Madreans believed that the heritage of the cabins and the gurgle of the stream were more valuable than the flood control that LACDA offered them. Finally, in May 1976, the city council resolved "not to proceed with any project plans for the Sierra Madre Wash" and recommended that "all affected members of the community investigate the benefits of the federal flood insurance program." That August the LACFCD asked the Army Corps to assign the Sierra Madre Wash project to inactive status.[35] The final piece of LACDA would not be built; for the first time since the 1930s, political conflict killed a project that engineers deemed technically sound.

In addition to the Sierra Madre Wash battle, environmentalists could point to a few other victories by the mid-1970s. As the rhetoric of Colonel Roper and the Flood Control District's *Biennial Reports* indicated, their activism had caught the attention of the flood controllers. As a result, the agencies began conducting more public hearings, held them in accessible places such as local high schools, and set the meetings' agendas to make sure citizens got to speak first, before private organizations and government representatives. The Flood Control District began spending hundreds of thousands of dollars annually to plant trees at its facilities, and it opened numerous recreation sites, including, in 1977, the LARIO trail, a twenty-mile bicycle and equestrian path along the Rio Hondo and the Los Angeles River from Whittier Narrows Dam to the Pacific Ocean. The Flood Control District and the Army Corps also began to explore a strategy called floodplain management, which integrated flood-control works with nonstructural measures such as zoning and urban planning in order to attain a "sensible, affordable, and environmentally compatible means that will provide adequate protection to lives and property."[36] Most visibly, the environmentalists blocked, delayed, or modified a few projects such as the Sierra Madre Wash improvement.

Despite these important but isolated successes, however, 1970s environmentalism did not genuinely change the goals of flood control or the process by which decisions were made. Pleas for parks and landscaping could be easily incorporated into the rhetoric of flood-control planners and even carried out in many cases without changing the overall shape of the institution. This is in fact what happened. Whenever environmentalism seriously challenged the program—for example, in Sierra Madre and other cases where environmental concerns did not easily fit into the allegedly best engineering schemes—the flood controllers balked. As Omer Hall's tirade during the Sierra Madre City Council meeting indicated, the district and the Army Corps would do things their way or not at all. Meanwhile, environmentalists' NIMBY (Not in My Back Yard) approach and emphasis on aesthetics limited their ability to formulate persuasive explanations for why such concerns should be incorporated. "I know that it is more expensive," the Sierra Club's Rosenthal pleaded with Colonel Roper at the 1972 LACDA hearing, "but I think there are increased values to having something that looked pleasing."[37] Many flood controllers considered such reasoning flimsy and sentimental. They agreed that things that were pleasing to look at were nice but not when those things interfered with hard cost and safety decisions. One Army Corps engineer, for example, acknowledged that environmental issues were serious problems, but he also insisted, "We cannot permit ourselves to yield to an emotional impulse that would make their cure the central purpose of our society. Nor is there any reason why we should feel guilty about the alterations which we have to make in the natural environment as we meet our water-related needs."[38] Despite the grassroots revolution and the flourishing culture of environmentalism in the 1970s, whenever things that looked pleasing weighed in against water-related needs, environmentalism usually came off sounding emotional and guilty. Consequently, in the Santa Clara River valley, which straddled Los Angeles County's northwestern border, and in many other sites of proposed LACDA expansion, the corps considered environmentally compatible floodplain management but often quickly discarded the option as too expensive.[39] Until flood-control opponents could expand their arguments beyond complaints about the aesthetics of their own neighborhoods, the inroads of environmentalism would remain confined to a few small projects and the pages of the *Biennial Reports.*

Fiscal crisis in 1978 further revealed the limits of the influence that environmentalism had attained over flood-control policy making. In June of that year, California taxpayers overwhelmingly approved Proposition 13, which drastically cut the state's property tax, the principal source of funding for local government. The new law doomed many public agencies, including the Flood Control District, which was largely responsible for operating and maintaining LACDA now that the Army Corps had completed construction. Almost entirely dependent on property taxes, the Flood Control District's budget plunged from thirty-seven million dollars to thirteen million, a 65 percent cutback. Eventually the agency disappeared entirely, its functions absorbed by the county's Department of Public Works in 1985. Before that happened, however, the district reassessed its responsibilities in order to determine where to spend its limited funds. In the new prioritization, floodplain management ranked ninth among thirteen district functions. Recreation and landscaping fell dead last. The district deferred all landscaping improvements in 1978 in order to finance the repair of structures damaged by storms earlier that year. Since 1969, the district had included a several-paragraph section on recreation, landscaping, and environmental enhancement in its *Biennial Report.* That section in the 1977–1979 report was one sentence long. It would never appear again in district reports.[40] It seemed that everyone agreed that parks and landscaping were nice, but whenever money got tight, they were the first things to go.

THE ENVIRONMENTAL CHALLENGE: THE RETURN OF FLOODS

The same 1978 storm that diverted Flood Control District resources from landscaping efforts also led engineers for the first time since the

1930s to question the technical adequacy of LACDA. Storms in February 1978 produced the most dramatic debris flows since 1934, and 1980 brought the series of storms that caused the Los Angeles River to top its banks at Wardlow Road. Together, the two disasters resulted in five hundred million dollars of damage and killed fifty-six people.[41]

The most remarkable feature of the storms was their small size relative to the havoc they wrought. Engineers had long expected that a great deluge might someday overwhelm the system, but these floods were well within LACDA's design capacity. To be sure, the environmental circumstances that accompanied the disasters had done plenty to prime southern California for catastrophe. Two hundred eight square miles of mountain watershed burned between 1975 and 1979, and during both the 1978 and 1980 storms, steady rains fell for extended periods, with the heaviest coming at the end.[42] Even considering these adverse conditions, however, the 1978 and 1980 storms were comparatively small events, much less severe, for example, than those that produced the floods of 1938 and 1969. Scientists rated the two recent storms as approximately twenty-five-year to forty-year events. "These floods," a scientist from the California Institute of Technology concluded, "were well within the range of frequencies for which the flood control systems have been designed. They were definitely not of disastrous proportions."[43]

The two storms nevertheless produced damage of disastrous proportions. Debris basins sent mudflows surging into neighborhoods. Levees crumbled while carrying less than their design flow. A desert lake swelled to several times its historical size. As a result, flood controllers began to reevaluate the assumptions about environment, engineering, and society that had guided the construction of the LACDA projects. Evidently, something more than the fury of the storms propelled the flow of water, something that gave engineers pause.

Throughout the twentieth century, flood control in Los Angeles had been predicated on assumptions about predictability. Both the federal and local programs proceeded from the premise that, given enough data, engineers could predict the amount of rain that would fall and the runoff that would result. Both programs furthermore relied on the assumption that, on the basis of these predictions, engineers could design structures to make the runoff flow according to human prescriptions. Finally, the programs demanded that engineers design flood-control systems that would accommodate future urban growth. These assumptions have not always been borne out in practice: engineers have been repeatedly surprised by the size and odd characteristics of floods; flood-control devices have not always worked exactly as designed; and flood controllers have frequently blamed flood damage on urban growth that created hazards before engineers could prevent them. Nevertheless, none of these facts had made predictability seem impossible during the decades of midcentury climatic quiescence. The limitations of prediction became dramatically apparent only in the storms of 1978 and 1980, which were not merely the results of freak weather. They were failures of a messy, nonlinear, chaotic urban ecosystem.

Foothill Flip Buckets

Rain, runoff, and flood-control designs turned out to be anything but predictable. In the San Gabriel foothills in 1978, meteorological quirks, urbanization, fluid dynamics, and the flood-control devices themselves interacted in unforeseen ways to multiply a twenty-five-year storm into a two-hundred-year disaster.[44]

Part of the problem stemmed from the fact that a twenty-five-year storm is not just a twenty-five-year storm. The statistical construct of a twenty-five-year storm is an abstraction of reality; in theory, centuries might elapse between twenty-five-year events, or Los Angeles might experience two in three years, as happened between 1978 and 1980. LACDA was designed to withstand at least a fifty-year storm. Although the terms are statistically logical, nature does not package its storms in discrete twenty-five and fifty-year units. Instead, each storm is an amalgam of varying intensities. That is, in a storm estimated as a twenty-five-year event, some hillsides might get only a drizzle while other slopes nearby might suffer a thousand-year drenching. Some places near Santa Barbara, to the northwest of Los Angeles, for example, suffered in 1978 rains of intensities that meteorologists expected them to experience on average once every three thousand years. During the same storm, a weather cell roared up the recently burned western slopes of the San Gabriel Mountains. Before it dissipated over the highest peaks, this turbulent weather pelted the watersheds above the foothill canyons with precipitation levels on the order of a fifty-year event, locally doubling the rating for the storm as a whole. As a result of the heavy rains and burned slopes, the debris production exceeded what engineers had anticipated.[45]

To make matters worse, urbanization and bad luck prevented the debris basins from performing as designed. By 1978, residential development had mushroomed around the mouth of Zachau Canyon above Sunland in the northern San Fernando Valley. The debris flowing out of the canyon during the storm rolled a six-foot boulder into the culvert under Seven Hills Drive. The plugged drain sent the debris cascading over the road and into a debris basin built in 1956 with an expected capacity sufficient to catch the muck from a fifty-year flood. But the muck rumbled right over the barrier as if it were not there and crashed into the neighborhood, damaging fifty homes.[46] Elsewhere, in Shields Canyon, above the La Cañada Valley, suburban development had crept uphill and leapfrogged over a protective debris basin in the 1960s. A second basin was later built above the residential area with a channel to divert flow around the homes and into the lower basin. As at Zachau Canyon, a boulder plugged the Shields channel and forced all of the flow into the neighborhood. The mud and water followed Pine Cone Road downhill until the street curved, at which point the debris thundered into homes and yards. The total amount of debris in the flow was not enough to fill the lower basin. Had the channel functioned, little or no damage would have occurred.[47]

Once besieged by weather cells, suburban development, and rolling boulders, the design of the flood-control works sometimes actually contributed to the destruction. Anticipating that the debris would flow at more or less regular rates and settle into the debris basins, engineers had designed the pits large enough to catch the expected amount of material. In reality, however, the masses flowed in surges that crested the dams at Zachau and other debris basins and continued downstream.[48] One Flood Control District engineer who found these fluid dynamics inexplicable characterized the failed debris basins as giant "flip buckets."[49] Just as twenty-five-year storms produced greater than twenty-five-year rains, debris basins with a hundred-thousand-cubic-yard capacity proved unable to catch a hundred thousand cubic yards of debris. With the most intense rains falling on the most easily eroded and thoroughly urbanized areas, and the debris basins behaving like flip buckets, the flood-protection system designed to handle a fifty- or hundred-year storm failed to contain a twenty-five-year event.

Crumpled Levees

Even well-designed structures operating below their design capacities behaved differently from what was intended. Upon first review, a flood-control channel seems to be a simple apparatus. It consists of two walls strong enough and high enough to keep a river flowing in between. That is what the Army Corps of Engineers intended when it designed its San

Jacinto River channel, which ran through San Jacinto, a town with a population just over five thousand at the eastern edge of the Greater Los Angeles metropolitan area. Local landowners had built earthen levees along the river and its tributaries beginning in the early 1900s, but these sandy walls had given way during numerous floods. To protect the town and the valuable agricultural land around it, the Army Corps installed state-of-the-art flood-control works in 1961.[50] The system consisted of a pair of eleven-foot-tall levees, each blanketed with gravel. Topped with quarried rock, this revetment extended ten feet below the surface of the stream bed. These rock-lined banks encased an earthen riverbed designed to carry flows of eighty-six thousand cubic feet per second, nearly double the largest San Jacinto River flood on record and sufficient, the corps expected, to withstand a 250-year storm.[51]

During the storms of February 1980, water rose in the channel. By late afternoon on 15 February, flood-fighting operations swung into high gear, as crews dumped rock to fill small voids that had opened in the levees. Finally, around two o'clock on the morning of the seventeenth, water swamped the bulldozers, forcing the crews to abandon their machines. For three days, the water continued rising and flowing faster. Around seven o'clock in the morning on the twenty-first, the levees crumbled and the river gushed through the gap. Walled off from the channel by the downstream levees, which had remained standing, the new river flowed parallel to its old channel for a short while, then turned down Main Street, storming through the city and spreading seven miles wide before eventually rejoining its historical flood plain and returning to the riverbed downstream from the levees. The city of San Jacinto and its vicinity sustained ten million dollars in damage before the army managed to repair the levee on the twenty-third.[52] At its peak, the river had flowed at twenty-five thousand cubic feet per second. The levees had failed at less than a third of their design flow.

Although engineers were not sure exactly how, the levee itself appeared to have contributed to its own failure. A team of Army Corps investigators determined that the river had eroded its own earthen bed to depths well beyond what the channel's designers had planned for. Once the bed was eroded enough to expose the lower reaches of the armored levee, the water ate away at the levee from underneath until the structure collapsed. Aggravating these conditions was the alignment of the channel. A tributary just upstream from the worst breach directed its waters into the main channel at an angle so that flows ran diagonally across the bed, hitting the opposite levee right at the spot that first broke.

Not only did this condition contribute to the erosion that undermined the levee; once the walls crumpled, the diagonal flow directed the water outward through the void and into the city. The Army Corps investigators traced additional breaches to the paving of upstream channels, which increased scouring of the downstream riverbed.[53]

J. David Rogers, a geotechnical engineer who also studied the event, suggested the collapse may have had something to do with bureaucracy as well. The design of the levee was fine, he said, but it called for construction with local materials that later turned out not to meet the Army Corps's design criteria. The flood-control bureaucracy, however, offered few options for changing the construction materials. Given the long procedure for procuring federal money, Rogers indicated, there was "no meaningful mechanism" by which a project could be changed once under construction "without someone's head rolling." To save money and to avoid a long and possibly embarrassing process of redesigning the levees, the builders used the easily eroded material available on-site.[54] Even when prediction and design worked out as planned, bureaucracy could lead to faulty construction and disaster.

Regardless of whether Rogers or the Army Corps team was right—and it is possible that both were—the San Jacinto levee was not a simple apparatus. Instead, it was a dynamic structure that interacted in complex and constantly changing ways with the water it carried. Even at moderate flow, small misalignments in the channel, construction of upstream levees, and inadequate depth of the revetment caused the water to undermine the levees. This physical structure, possibly with some assistance from arcane design and construction procedures, led the levee walls to fail in 1980 and turned a twenty-five-year flood running at one-third of the system's design capacity into a catastrophe.

The Lake That Doubled in Size

Further confounding the predictability of flood control has been society's habit of placing valuable objects in harm's way. Dry-period development of floodplains contributed to the 1914 and 1934 floods, a pattern that repeated in 1980 at the desert resort of Lake Elsinore.

Sixty-five miles southeast of Los Angeles, Lake Elsinore sat more than twelve hundred feet above sea level, surrounded by mountains. In most years, the five-by-two-mile lake was the terminus of the San Jacinto drainage basin and collected the thirteen inches of precipitation that fell annually on the lake's surface as well as the twenty-five or more inches that showered the mountain slopes above it. Like elsewhere in southern California, these rainfall totals fluctuated considerably from year to year, and during wetter seasons, the lake splashed through an overflow channel into Temescal Wash and eventually out to the ocean through the Santa Ana River. At maximum capacity, the surface of the lake lay at an elevation of 1,260 feet and covered an area of sixty-two hundred acres. For most of the century, however, the surface area averaged only thirty-eight hundred acres, and it never overflowed into Temescal Wash in the sixty-four years after 1916.[55]

During this period, Lake Elsinore became a favorite playground for southern California urbanites. The same rising affluence and congestion that sparked the national environmental movement in the postwar years also fueled a recreation boom at the lake. Although it was almost completely dry during the droughts of the 1950s, the lake rose by the 1960s, and water-skiers, motorboaters, campers, parachutists, and hang gliders flocked to Lake Elsinore's scenic views and calm waters. Vacation homes, trailer parks, campgrounds, businesses, and resorts ringed the shores by the 1970s. With the water level never surpassing 1,248 feet in elevation during these years, development even crept into the old lake bed.[56] The 1916 flood had all but faded from memory.

Flooding rushed back into the minds of southern Californians in late February 1980. The elevation of the lake stood at 1,247 feet above sea level on 13 February, when the first rains began to fall. By the twenty-third, thirteen inches of rain, an entire year's worth, had fallen, doubling the volume of the lake and raising its surface elevation to 1,259 feet, one foot below its overflow level. Even after the rain stopped, the lake continued to swell with runoff from the surrounding mountains. During the six decades since the lake had last overflowed, vegetation and debris had clogged the Temescal Wash channel, raising its bed above the outlet from the lake. So the lake kept rising. Instead of filling the wash, water overflowed the streets, homes, and businesses that lined the shores. The Army Corps and its contractors worked twenty-four hours a day to unclog Temescal Wash until 7 March, when the lake finally flowed into its outlet. Its elevation was 1,264 feet. Before it receded, the lake flooded three hundred houses and numerous mobile homes and businesses. Two thousand residents had been displaced, and the damage toll approached fifty million dollars. Strict enforcement of zoning laws already in effect would have prevented the devastation, but in the heat of the tourism boom there had been no political will to guard against a threat that nobody imagined.[57] Just as in 1914, long climatic cycles, short human memories, and rapid development transformed a comparatively small meteorological event into a big disaster.

Reassessment

The 1978 and 1980 floods marked a turning point in southern California flood-control planning. Since the 1940s, flood-control efforts had been guided by a fundamental belief that the LACDA project could control the rivers. Most observers admitted that perhaps the most extreme storms would overwhelm the system, but those were seen as the exception. Disasters like that of 1938, planners believed, were no longer possible. After the moderate, but still destructive, storms of 1978 and 1980, however, flood controllers confronted the possibility that perhaps something less than an exceptional storm could do considerable damage—even if structures were built according to design and in the appropriate locations. As urban development, quirky meteorological phenomena, short human memory, recreation booms, engineering bureaucracy, and even flood control itself interacted to generate a catastrophe out of proportion to the external forces that triggered it, flood controllers began to reassess their assumptions and strategies. When they did, many reached the same conclusion as the California Institute of Technology's Norman Brooks: the physical structures of LACDA were "necessary but not sufficient."[58]

The National Academy of Sciences' National Research Council (NRC) began that reassessment just months after the flood of 1980. After the 1980 storms, the council, together with the Environmental Quality Laboratory at Caltech, convened a symposium in Pasadena to discuss the storms of 1978 and 1980. The group that assembled at Caltech on 17–18 September included state, federal, and local engineers, geologists, and meteorologists, as well as a number of university scientists. This collection of experts resembled the technocratic gatherings that had followed previous flood disasters, but this group came to some strikingly untechnocratic conclusions.[59] The five hundred pages of proceedings published in 1982 unmasked a flood-control program short on technical knowledge and overly dependent on physical structures.

A recurrent theme of the symposium was the inadequacy of existing information about floods. Although the Flood Control District and the army's Los Angeles District together constituted one of the greatest storehouses of flood-control knowledge in the world, the 1978 and 1980 storms revealed how little flood controllers really knew. Statements of uncertainty peppered the pages of the NRC's proceedings. "The old rule of thumb that scour depth equals flow depth is incorrect," acknowledged the investigators of the San Jacinto levees.[60] The foothill debris flows inspired a Flood Control District official to conclude that "the previous understanding of sediment deposition, based on experience, was inadequate."[61] The symposium's moderator, Norman Brooks, observed that flood controllers still needed to know a lot more about long-range weather forecasting, occurrence of weather cells, effects of urbanization, optimum reservoir management, levee design, mechanics of mudflows, behavior of small canyon watersheds, controlled burning, bank and bed erosion, nonstructural strategies, and even basics like mapping.[62] Not even the electronic stream and rain gauges, satellite meteorology, and computer modeling that flood controllers had come to rely on in recent years had enabled them to master the dynamics of flowing water.

Worse yet, it appeared that some pieces of precise information might not be possible to gather at all. Although the scientists freely tossed around terms such as *twenty-five-year storm,* they admitted that such estimations of frequency were difficult to make. Their data period was too short, they said. Different measurement sites yielded different information. And urbanization was constantly changing conditions, so that what looked like a twenty-five-year flood on the basis of past experience might in fact become considerably more frequent in the future.[63] The Army Corps's San Jacinto investigation team noted that in wide, meandering streams with variable flows, the points where levees would suffer the greatest stress "may be indeterminate."[64] The flow of water carrying mud and debris presented even bigger problems. "We can identify areas that are prone to landslides and mudflows," Brooks said, "but we do not have the ability to predict just when any particular slide might occur."[65] Even knowing that, however, would be only partly useful, because the fluid mechanics of flowing debris are so complex. Hydraulic equations that modeled water were inadequate for modeling debris flows, because they were based on different assumptions about viscosity and density. Nor did stream gauges provide much reliable data, because they were often buried by the very events they were supposed to measure. Velocities of flow were hard to calculate, because the average-current meters could not account for the surges that had turned the debris basins into flip buckets. All this led Flood Control District officials to conclude that the velocity, depth, and kinetic energy of debris flows were much higher than they had suspected. By how much, however, no one was sure. "Reliable postflood measurements of sediments that passed debris basins," lamented one Flood Control District official, "are nearly impossible to make."[66] As the scientists confronted the sometimes inexplicable events of the two storms, their Progressive Era confidence in the essential rationality of nature eroded just a bit.

So did their confidence in LACDA itself. Many papers at the symposium emphasized the need for nonstructural measures to complement the physical works. While some of the damage in 1980 was due to an "unusual event," a state Department of Water Resources official observed, "most of the damage could be attributed to a lack of foresight." Adding that "the floods of 1978 and 1980 have clearly demonstrated the need to implement floodplain management," he called for better enforcement of zoning and building permits. The failures of the structures in the 1978 and 1980 floods inspired such assessments, but there were other reasons too. By the early 1980s, the cost of public works had skyrocketed, and the national economic recession of the 1970s had ended the long period of fiscal prosperity that postwar government agencies had enjoyed. The environmental movement was challenging every new structure.[67] "New public works construction," one Flood Control District official contended, "will probably receive relatively low priority within the next generation. Consequently, people should not expect that even the existing flood and sediment hazards can be cured," much less the new hazards that urbanization was creating daily.[68] Given this economic and political climate, future flood protection would have to come from nonstructural measures. The public, the official said, would have to abandon the "viewpoint that the U.S. Army Corps of Engineers or the local flood control district will correct any problems caused by unwise development."[69] Finally, nonstructural measures appeared to work: strict building codes enforced on all new hillside construction in the city of Los Angeles since the 1960s had spared that city's foothill residents much of the damage suffered by other areas not subject to the codes.[70] The types of nontechnical strategies that were virtually unimaginable to the flood-control establishment in the 1930s now, under different historical circumstances, seemed imperative.

Were the technocrats admitting defeat? Hardly, but the 1978 and 1980 storms, which did damage so out of proportion to their size, shook their confidence. In particular, the debris on the levee near Wardlow Road spurred the Army Corps to intensify an already-underway LACDA review, the results of which were alarming. LACDA's designers in the 1940s had expected the region to remain at least partly agricultural and therefore capable of absorbing some of the rainfall that fell on it. Between 1940 and 1980, however, the county's population grew 270 percent, and urbanization made nearly the entire surface of the Los Angeles Basin impervious to water.[71] By 1980, when the orange groves and walnut orchards had all but disappeared, urban runoff had increased 25 percent, and LACDA had grown inadequate to protect the strip malls, subdivisions, and freeways that had replaced them.[72] Making matters worse, the two thousand miles of underground storm drains implanted to protect all this development from local flooding concentrated and sped runoff into the main river channels, producing peak flows that were considerably higher than expected.[73] Thus, separate components of the flood-control system were linked in ways LACDA's designers had not anticipated, and the result was that the fifty- to hundred-year flood protection the corps thought it had secured was downgraded to twenty-five- to forty-year protection. In a hundred-year storm, the corps predicted, the levees would fail, inundating an eighty-two-square-mile section adjacent to the Los Angeles River and causing more than two billion dollars of damage. Barely a decade had passed since its completion, but LACDA was already obsolete.

This was a potentially expensive discovery. Under the National Flood Disaster Prevention Act of 1973, holders of mortgages backed by the federal government (some 95 percent of all mortgages) are required to purchase federal flood-control insurance if their property is in a hundred-year floodplain.[74] The corps's 1986 announcement that a half million people and 142,000 structures occupied the lower Los Angeles River's hundred-year floodplain meant that tens of thousands of homeowners in the predominantly working-class and low-income neighborhoods in the threatened area might have to purchase policies costing as much as $281 per year for $100,000 of insurance. Although at first the requirement was laxly enforced, the Federal Emergency Management Agency (FEMA), which sets flood-insurance rates and determines who must purchase it, began redrawing its own maps and announced in early 1998 not only that residents in floodplains were now subject to the flood insurance mandates but also that the agency was doubling the cost of such insurance. Twelve hundred angry people turned out to assail officials at a FEMA open house in March, and hundreds more protested the rate hikes at town meetings and public demonstrations during the next few months.[75] In July, just four days before the new rates were to become effective, FEMA backed off. Property owners in the floodplain would still be required to purchase insurance, but they would be allowed to do so at the old rates. Certainly the groundswell of public

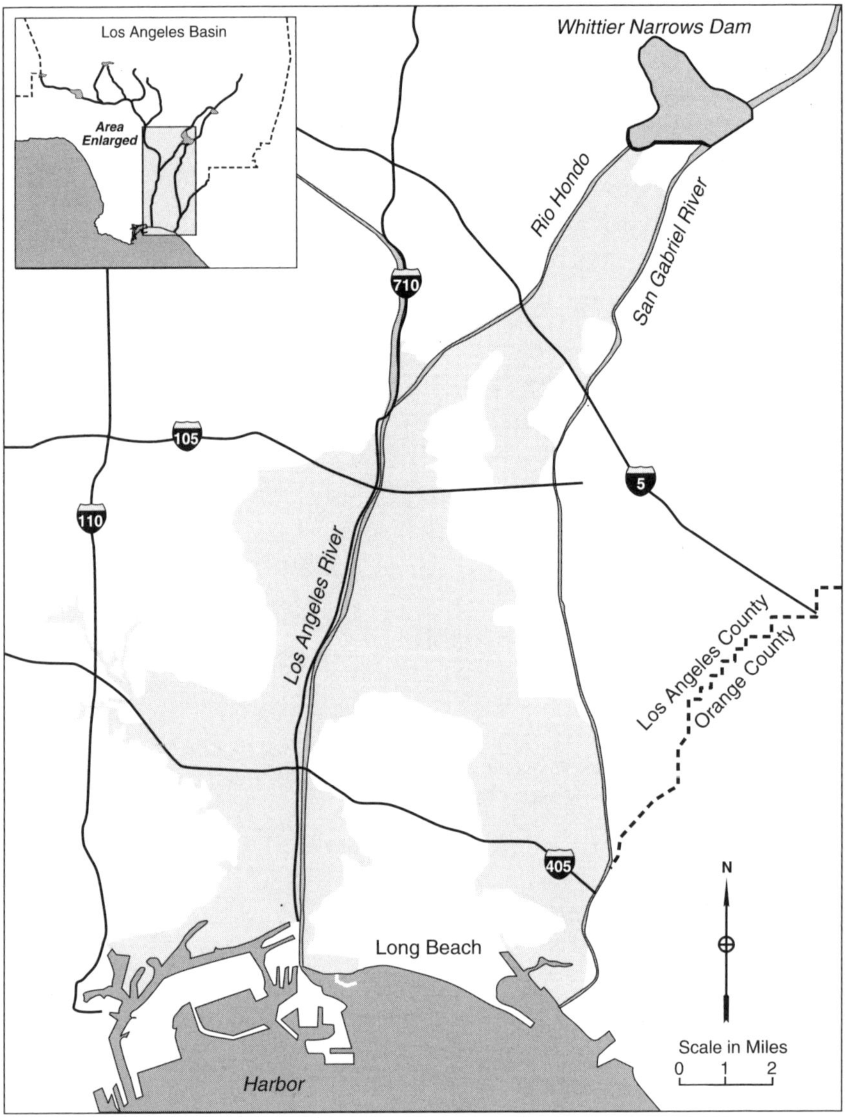

Map 8. The shaded portions indicate the areas that were thought to be subject to inundation in a hundred-year flood, 1992. The coastal plain area previously traversed by the braided channels in the 1890s was still threatened with overflow a century later, despite having a very different geography. Compare with map 2 (page 21), which depicts nearly the same area one hundred years earlier. (Map by Carol Marander.)

protest and the fact that the floodplain crossed parts of six congressional districts influenced FEMA's last-minute concession, but so did something else.[76] In June 1998, Congress had allocated funding for yet another plan to tame the rivers.

As a result of its LACDA review, the Army Corps proposed redesigning parts of the channels along the Rio Hondo and the lower Los Angeles River. As they had done for seven decades, the engineers surveyed the landscape—this time from computer screens—recalculating expected flood flows, channel volumes, and levee strengths. The new plan called for widening channels, modifying bridges, and, most important, constructing parapet walls, two to eight feet tall, on the levee tops along twenty-one miles of channels. The walls, along with the other modifications, would effectively enlarge the channels to provide hundred-year flood protection and enable FEMA to remove the insurance requirements for the adjacent lands. Congress approved the $364 million plan in 1990 but dragged its feet on the funding. The project looked as if it might take as long as twenty years to complete—with the floodplain homeowners each shelling out hundreds of dollars in insurance for every year it was delayed, a burden that one critic labeled a "federal working-class tax." As the flood-insurance opponents grew more boisterous in 1998, however, Congress increased funding to speed up the project, and it was completed at the end of 2001, five years ahead of schedule and $150 million under budget.[77] In 2002, it won one of the nation's most prestigious public works awards, the American Public Works Association's Project of the Year.

The project was simultaneously a culmination of and a departure from a century of technocratic flood control in Los Angeles. On one hand, this latest chapter had much in common with its predecessors: rapid urban development followed by a flood that warned of the inadequate protection, technical experts crunching data to produce a plan with the best possible cost-benefit ratio, flood control effected with bulldozers and concrete, and lofty promises of future protection. On the other hand, much was different. The Army Corps received considerable public input through letters, media attention, and hearings. It undertook the review and the construction in a legal context sensitive to environmental concerns, and, accordingly, it conducted an extensive environmental impact investigation. Also in contrast to midcentury flood control, the project generated considerable opposition across the Los Angeles area, and the corps had to fend off a lawsuit by environmental groups that objected that the parapet walls further degraded the blighted rivers and constituted a missed opportunity for more environmentally friendly flood control. Despite all this opposition, however, the project survived. The political pressures for protecting urban development—especially the mortgages of endangered homeowners and their lending institutions—were simply too great. "I hate it," said supervisor Gloria Molina, who voted for the plan in 1996 and who represented a district that lay partly in the newly identified floodplain, "but I have no choice."[78]

Although it appeared that the technocrats had gotten their way yet again, this first large controversy since the Whittier Narrows conflict suggested that flood control might have to be done differently in the future. The 1978 and 1980 storms had cast into doubt the efficacy of controlling water with bulldozers and concrete, and environmentalists were demanding a role in flood-control policy making. In this new political and ecological context, the question arose for the first time since 1914: How shall we try to control floods?

A GREENWAY FROM THE MOUNTAINS TO THE SEA

This meeting was a little louder than usual.[79] In the spring of 1990, the Los Angeles mayor, Tom Bradley, commissioned the Los Angeles River Task Force to "articulate a vision for the future of the river."[80] For the first time ever, the task force gathered engineers, environmentalists, city planners, politicians, artists, business leaders, recreation enthusiasts, and other concerned citizens to discuss issues affecting river management. The broad range of viewpoints on the task force made the first few meetings quite lively. Possibilities were limitless, and discussion of any one topic became difficult without confronting the many constraints against the issues it involved.[81] Eventually, someone suggested holding a brainstorming session at which the task force would generate as many ideas as possible without editing or passing judgment. The resulting list would establish the basis for organizing and assessing the rest of the task force's agenda.

On 25 October, the task force gathered for such a meeting. The facilitator, Peg Henderson, encouraged all participants to voice any idea they could imagine, without regard to feasibility. For the next hour, members broached issues related to flood protection, nature preservation, water quality, aesthetics, recreation, transportation, commerce, education, land use, interagency cooperation, planning, design, construction, acquisition, funding, maintenance, liability, and security. "Restore the river's natural habitat," one person suggested. "Create detention basins to control flooding," said others. "Take down fences." "Put fences back up." Build a "park from Hanson Dam to Long Beach."[82] The task force members wanted flood control, native plant species, graffiti prevention, equestrian trails, river cleanups, educational programs, low-income housing, and a monorail. Henderson and the mayor's office staff took all the ideas and organized them into topical and time-frame categories. By the time the meeting ended, the collaborators had generated 148 suggestions that would form the basis of the task force's deliberations and recommendations to the city.[83] In 1992 its official report recommended taking steps to meet flood-control needs and at the same time restore natural ecosystems, improve water quality, enhance riparian beauty, maximize public use of the river, develop alternative transportation options, and build public awareness of the river.[84]

As the meeting indicated, by the 1990s, technocratic flood control in southern California gave signs that it was undergoing fundamental transformation. It is hard to imagine the Army Corps and Flood Control District engineers in the 1950s inviting the public for a brainstorming session. It is even harder hard to imagine that any midcentury planning procedure would have produced the wide variety of public and environmental goals that the task force eventually recommended. As the task force struggled to integrate so many river uses and to involve so many river users, flood control in southern California faced its most radical possibility for change since its institutionalization in 1914.

The first step in this attempted reinstitutionalization occurred in 1985. Just as the Army Corps was coming to the conclusion that LACDA was inadequate, so were environmentalists, though for very different reasons. One morning in 1985, the poet Lewis MacAdams and three of his artist friends met for an early cup of coffee at an old dairy building on Vignes Street in Los Angeles. From this rendezvous they headed for the Los Angeles River. Clipping the Flood Control District's fence with a pair of wire cutters, they slipped into the forbidden territory and scrambled along the concrete faces of the riverbed. Freight trains rumbled along both banks of the stream. Traffic whizzed over two freeway overpasses. Jackhammering crews were ripping up pavement. It smelled industrial. The explorers, however, imagined the spot as a place of stately herons, darting steelheads, and marshy thickets. "We asked the river if we could speak for it in the human realm," MacAdams later recalled. "We didn't hear it say no."[85]

From this mix of art, civil disobedience, and outdoor adventuring, Friends of the Los Angeles River (FoLAR) was born. The founder, MacAdams, envisioned the new environmental organization as a "forty-year art project," a long-term term vision for reinventing—not just restoring—the river. Although MacAdams believed that FoLAR's work would not be done until the sycamores and yellow-billed cuckoos returned to the river, he deliberately conceived the organization's work as artifice, not nature restoration.[86] "To claim that the Los Angeles River has to be a wild river is absurd," he said in a 1993 interview. "You won't see the river be 'natural.'"[87]

Instead, FoLAR aimed to improve the existing artifice. In FoLAR's vision for the river, water would trickle over earthen low-flow riverbeds. Marshes and small ponds would abut the channel in the widened bottom of the bed, and riparian vegetation would supplant the concrete covering the sides of the banks. A pedestrian path would run somewhere above the two-year flood level. The tops of the levees, where Flood Control District fencing walled off utility roads, would become sylvan park lands with bicycle paths, athletic facilities, and picnic grounds. In this enlarged and greened channel, the river could rise even into the park land without doing great damage. The earthen bottom and vegetation of this multilevel channel would slow and spread the waters, allowing them to deposit their silts and sink into the ground.[88] FoLAR's flood-control channel promised to provide parks to inner-city youth, water-quality enhancement, educational opportunities, recreational and commuter bicycle transportation corridors, and habitat for the sycamore and yellow-billed cuckoo. Whereas groups like Sierra Madre's Save Our Stream had reacted against individual projects, FoLAR focused on creating a long-term vision for the entire watershed. As MacAdams explained, FoLAR labored "in service to an *idea:* creating a Los Angeles River greenway from the mountains to the sea."[89]

At first, FoLAR's biggest challenge lay in persuading the public that the river existed at all, for most southern Californians were only dimly aware that those big swaths of concrete were rivers. A 1978 article called the Los Angeles River "A Big Joke, a Killer in Concrete, a 50-Mile-Long Ditch," and in 1985, a writer who claimed to have lived in southern California for ten years and crossed the Los Angeles River fifteen thousand times without ever noticing it wrote a series of newspapers columns about the "alleged Los Angeles River." In fact, it became something of a cliché of Los Angeles River journalism for authors to confess how long they lived in Los Angeles before they knew there was a river.[90]

In drawing attention to the river's existence, FoLAR was greatly aided by a local state assembly representative's wacky 1990 scheme to turn the paved bed of the Los Angeles River into a freeway. Although engineers dismissed the plan as a perilous folly and even the assemblyman himself seemed to lose interest after his failed effort to win a Los Angeles mayoral election, his stunt left a lasting impact by reminding southern Californians that rivers did indeed lie under all the concrete.

Building on this burst of publicity, FoLAR sponsored plays, conferences, hikes, neighborhood meetings, tree plantings, art exhibits, and other activities to educate the public about the rivers. Most popular of all were the river cleanups. These events attracted hundreds of people annually, mostly from the ethnically diverse neighborhoods northeast of downtown, to remove trash and graffiti from the riverbeds. FoLAR hired high school bands to entertain the cleaners, and *folklórico* troupes danced on the levees.[91] FoLAR also held public conferences, lobbied local politicians, hired engineering and landscaping firms to design river-greening plans, threatened lawsuits to block projects, and helped found the Los Angeles River Center. After sprouting from origins in riparian civil disobedience, by the mid-1990s FoLAR boasted a board of directors, a technical advisory board, and a membership roll of several thousand names. More important, it had allies, as a host of other organizations—Trust for Public Lands, Northeast Trees, Tree People, the Sierra Club, the Santa Monica Mountains Conservancy, Occidental College, the Los Angeles and San Gabriel Rivers Watershed Council, and many others—joined the efforts to improve southern California's rivers.[92] "I have to hand it to Lewis," said one Public Works Department official in 1994. "He put the river on the map."[93]

By getting the river onto southern Californians' mental maps, FoLAR also got the attention of policy makers. Mayor Bradley appointed the brainstorming Los Angeles River Task Force in 1990, and subsequently, candidates for public office routinely made the river an issue in their campaigns. In July 1991 the Board of Supervisors voted unanimously to direct the county's Public Works, Parks and Recreation, and Regional Planning departments to "coordinate all interested public and private parties in the planning, financing and implementation efforts of a Master Plan for the Los Angeles River."[94] To carry out this task, the departments formed an advisory committee composed of engineers, environmentalists, artists, landscape architects, educators, museum employees, bicycle club members, and numerous federal, state, and local government officials. Not only were these planners diverse in their professional backgrounds, but also they were broadminded in their visions of urban river management. For decades southern Californians had feared the Los Angeles River as "a killer encased in a concrete strait jacket," and viewed its paved channel mechanistically as "a useful contraption," but the authors of the Los Angeles River Master Plan, in contrast, called the river "a complex resource," a "community amenity," and an "urban treasure."[95] They believed that the river, which flowed through thirteen municipalities and nine Los Angeles City Council districts, past mansions, subdivisions, and shacks, and through neighborhoods where almost every language under the sun was spoken, held "the potential to be a significant link between people and neighborhoods."[96] By the early 1990s, then, Los Angeles environmentalists could no longer be placated with *Biennial Report* rhetoric, isolated project modifications, and a little landscaping. Instead they insisted on new policies, new policy makers, and a new policy-making process; in short, they sought to reinvent the entire institution of Los Angeles flood control.

By the time the county released the Master Plan in 1996, FoLAR and other groups had generated plenty of ideas, energy, and attention, but no policy changes. The mayor's task force had no authority to do anything other than make recommendations, and even the Master Plan was primarily an advisory document. Some environmentalists attacked it as toothless. Some engineers considered it chimerical. One geographer characterized it as "more of a master suggestion than a master plan."[97] Meanwhile, FoLAR and its allies lost the battle to block the construction of the parapet walls of the Army Corps's LACDA upgrade. And the Army Corps even refused to spend $750,000 of congressional funding for recreation on the Los Angeles River.[98] Despite more than a decade of successful environmentalist efforts to draw public attention to the river and build public support for alternative river-management strategies, the flood-control technocracy seemed more resistant to nontechnical public input than ever. "We don't play politics," one engineer snapped in 1994. "We're engineers."[99] FoLAR's ability to play a new brand of politics, however, would soon deal the flood-control technocracy its biggest challenge.

One night in 1994, MacAdams and FoLAR board member Martin Schlageter ventured into territory that would have been unfamiliar to many environmentalists twenty years before. They came that night to St. Anne's Church, in Frogtown, an ethnically mixed working-class neighborhood northeast of downtown along the Los Angeles River, to pitch FoLAR's river-greening plans to the Vietnamese, Chinese, Filipino, and Latino residents. Specifically, MacAdams and Schlageter wanted reaction to a plan for building a park and flood-control retention basin at Taylor Yard, a two-hundred-acre abandoned Southern Pacific rail facility that was one of the largest remaining parcels of undeveloped land along the river and one of the neighborhood's biggest eyesores and health hazards. But northeastern neighborhoods had learned to be leery of such redevelopment schemes. In 1933 Chinatown was relocated to make way for construction of Union Station. In the 1950s a poor but vibrant Latino neighborhood in Chavez Ravine was sacrificed to build Dodger Stadium. Over the years, the tangle of freeways that crossed the area displaced still more homes and businesses. Remembering their long history of suffering at the hands of those who proposed urban redevelopment, the residents were understandably less than welcoming to these latest visionaries. "What developer do you represent?" they asked MacAdams and Schlageter. "What kind of secret deals?" "That railroad land is poisoned." "Who paid you off?" But the two environmentalists were undaunted. "What is your dream for this land?" MacAdams asked. "Trees!" came one response. "A museum!" came another. "Nature lessons for the kids!" "Soccer!" "A merry-go-round," chimed one old man. MacAdams could not scribble fast enough to keep up with their dreams.[100]

At the St. Anne's Church gathering and numerous other similar meetings, FoLAR representatives encountered a strand of grassroots organizing that had historically been parallel to but separate from their own brand of activism. Growing out of the same grassroots revolution that had inspired environmentalism, community groups that organized people, especially in neighborhoods and churches, around a variety of social justice issues proliferated in the 1960s and 1970s. Community organizers nationwide began increasingly to take on environmental issues after 1982, when poor African Americans in rural North Carolina challenged the federal Environmental Protection Agency and its cadre of scientists who recommended dumping soil contaminated with polychlorinated biphenyls (PCBs) into a local landfill. By the 1980s, southern California was a hotbed of such environmental justice activities, with groups such as Concerned Citizens of South Central Los Angeles, Mothers of East Los Angeles, and the Labor/Community Strategy Center, all of which worked against the disproportionate exposure of low-income ethnic and racial minority neighborhoods to air pollution. Penny Newman, a concerned mother from Glen Avon, on the eastern fringes of the metropolitan area, organized her community after the 1980 storms flushed hazardous waste from a nearby dump into the town. Newman went on to become an organizer for Citizen's Clearinghouse for Hazardous Wastes, a national organization dedicated to empowering non-expert citizens to participate in the highly technical policy making regarding toxins. To Newman, who once was admonished by industry representatives to "get beyond emotion and argue about technologies," the lessons of environmental justice and community organizing were clear: "We are the power. We are the experts. We are the ones who have watched our community devastated."[101]

Such community organizing efforts shared much with their more mainstream environmentalist counterparts. They resisted expert-driven policy making. They insisted that citizens have a say in policy making that would affect them. And they drew on moral and aesthetic arguments, not just technical ones. But community organizers and environmentalists had not often cooperated. Chi Mui, an activist in Los Angeles's Chinatown in the 1980s, for example, described himself as an "inner-city environmentalist" but said he had never worked closely with environmental groups. He knew they shared his interest in open space, clean air, and clean water, but those groups rarely came into the inner city, and so, he said, "I couldn't quite relate."[102] Across the nation, other activists expressed even stronger alienation, as a group of nonwhite organizers expressed in a 1990 letter attacking the "racism and 'whiteness'" of several prominent national environmental organizations.[103] As the environmentalists discovered at the St. Anne's meeting, inner-city residents were not, for the most part, interested in restoring steelhead, sycamores, cuckoos, or any of the other trappings of pristine nature; they wanted things that would improve the quality of life in the city—like merry-go-rounds and soccer fields. The St. Anne's meeting ended in a draw, with the residents voicing their wishes and the environmentalists agreeing to take the plan back to FoLAR's drawing board. But here was a crossroads for both groups: the path of continued mutual suspicion or the possibility of a powerful alliance. They chose alliance. Over the next few years, community activists and environmentalists would learn to cooperate, and that alliance would shortly generate what appears to be one of the greatest successes in both community redevelopment and environmental preservation that Los Angeles has ever seen: the rehabilitation of the Cornfield.

The Cornfield, also known as Chinatown Yard, is a fifty-acre parcel of land that stretches northeast from downtown, from the old pueblo where the city was founded in 1781 to the Los Angeles River. Taking its name from the corn grown there in the nineteenth century, it is a site on which one can trace the history of Los Angeles. The first Europeans to come to southern California probably camped near the site in 1769, and the Zanja Madre, the ditch that supplied the young town's water, ran from the river across the site to the pueblo. The founders and their descendents farmed the land for one hundred years until the railroads brought the first urban boom to Los Angeles in the 1870s, at which point the Southern Pacific Railroad began using the site as a rail yard until abandoning it when the freight industry moved inland in the second half of the twentieth century. By the 1990s, the Cornfield was unused like Taylor Yard just upriver, abandoned, and highly toxic, considered a blight by the twenty-five thousand mostly Chinese and Latino people who lived in neighboring Chinatown. It was also the site of grand but competing visions for urban redevelopment. Beginning in the late 1980s, residents of this long-impoverished area sought city aid to redevelop the site, but their pleas were mostly ignored. Then in 1999, Majestic Realty Corporation, the city's largest developer and the builder of the downtown Staples Center sports arena, proposed to purchase the site from the railroad. Los Angeles mayor Richard Riordan helped the company secure a number of incentives, including a twelve-million-dollar grant from the U.S. Department of Housing and Urban Development (HUD), to convert the site into an industrial park of warehouses, food-processing plants, and garment factories. The eighty-million-dollar project promised to turn unproductive land to profit and to create a thousand jobs, and it gained the support of the Chinatown Chamber of Commerce and numerous Los Angeles politicians, including city council member Mike Hernandez, in whose district the Cornfield lay. Majestic's plan was typical of the policy-making style that had characterized so much of Los Angeles's development during the century, including flood control. Decisions were made in the service of economic development by an alliance of experts in industry and in local and federal government. Nobody asked the people who lived there what they wanted.

Many Chinatown residents, however, did not want the warehouses. One of these was Chi Mui. A Chinese immigrant, Mui had been organizing in Chinatown for nearly twenty years when he first heard of Majestic's plan, and his long involvement in the community had alerted him to his neighborhood's pressing need for open space. There was nothing green in Chinatown. It was overcrowded. Kids played soccer on cement fields. The nearest middle school was two and a half miles away—across the river and two freeways. The proposed warehouses would solve none of these problems. They were, in his eyes, a deception. "History has taught the leaders of Chinatown that huge projects like this will basically kill Chinatown," he told a journalist in the spring of 2000, perhaps recalling Union Station. "They won't bring a thousand jobs. They'll only bring freight trucks that will drive the tourists away. We have enough warehouses, but parks?"[104]

Parks interested MacAdams as well. The Cornfield had long appealed to FoLAR as a green space adjacent to the river that could serve as an overflow site to detain floodwaters and possibly enable the future removal of some concrete from the bed of the Los Angeles River. After the landscape architect Arthur Golding designed a park and neighborhood development plan for the site, MacAdams approached Mui in 1999 and told him of his and Golding's visions. Mui appreciated that MacAdams had not "come in here with an agenda," but rather had come to listen and to work for a solution that benefited the community. MacAdams, for his part, thought Mui was a "brilliant organizer."[105] Together with Golding, Mui and MacAdams united two historically separate strands of the grassroots reaction against expert-driven policy making: environmentalism and community organizing. River greening and urban justice proved to be a potent mix.

As any good neighborhood organizer would, the three began by knocking on doors and talking to people in the community about their vision. Tapping Mui's extensive community connections, they won the support of the Chinese Consolidated Benevolent Association and the Chinese American Citizens Alliance, local branches of national organizations with decades of experience defending the interests of Chinatown residents. They also met with senior citizens groups and neighborhood organizations such as the Alpine Neighborhood Association, which worked as a liaison to the police, trash collection, and other city services. That was another tactic borrowed from community organizing: reliance on existing leadership and institutions within the community. The backing of the Chinese Consolidated Benevolent Association and other groups proved crucial in winning community support and shielding the organizers from the charge of being outside white folks advancing middle-class goals without truly understanding what the community needed. The dogged organizing efforts produced the Chinatown Yard Alliance, a multiethnic and multiclass coalition of more than thirty organizations ranging from the Sierra Club to the Chinese Consolidated

Benevolent Association to Mothers of East Los Angeles to the Latino Urban Forum to local soccer clubs. Meeting regularly over dim sum at a Chinatown restaurant, the coalition members hammered out mutually appealing alternatives to Majestic's industrial development proposal and began trying to persuade officials to consider it. They won the support of each of the 2001 Los Angeles mayoral candidates and gained a public relations boost when two amateur archaeologists unearthed at the Cornfield a historic portion of the Zanja Madre in the summer of 2000. But despite these small victories, the coalition was unable to stop the Majestic project from sailing through the city planning offices charged with reviewing it.[106]

At that point, the Chinatown Yard Alliance's environmentalist roots became an asset. Because of FoLAR's participation, the alliance enjoyed the backing of several established national environmental organizations and employed a superb legal team that exploited a crucial Majestic error. In the summer of 2000, when the company persuaded a city planning commission to approve the warehouse development without a full environmental impact report (EIR), the alliance lawyers filed suit in federal court and used their connections to HUD secretary Andrew Cuomo to get him to withhold Majestic's grant pending the completion of a full EIR. That killed it. As one Majestic vice president put it, "In order for the development to be successful . . . financial assistance was required." Facing both an EIR and a costly lawsuit, Majestic agreed to sell its option on the Cornfield to the Trust for Public Land, a national parks conservancy organization, for thirty million dollars (turning an eleven-million-dollar profit over its own initial outlay for the site). In 2001, the state purchased the land from the trust, using funds from Proposition 12, a $2.1 billion bond measure for parks California voters had approved in March 2000. By early 2002, the newly established Cornfield State Park Advisory Committee began to explore possibilities for the site including a state park, a middle school, two soccer fields, a Chinese cultural center, an institute for Chinese philosophy and martial arts, bicycle paths, a museum dedicated to the city's early water system, a canal that would flow along the old course of the Zanja Madre, and a lake that would double as a flood diversion receptacle.[107] The first large patch of the Greenway was to become a reality.

Environmentalism had come a long way. Environmentalists who met with working-class urbanites in Frogtown churches and Chinatown restaurants had expanded their vision considerably since their predecessors tried to exclude blacks from Sierra Club membership in the 1950s.

Along the way, they also developed a new brand of environmentalism. As Schlageter put it, "The Los Angeles River isn't Yosemite."[108] Environmentalists in Los Angeles at the beginning of the twenty-first century have devoted themselves less to preserving wild Utah canyons than their predecessors did, but they have poured themselves into improving the places where southern Californians live. They have been concerned less with the rustic character of mountain neighborhoods and more with health and safety in the most polluted ones. They have spent less effort preserving patches of wilderness in Sierra Madre Canyon and more time lobbying for soccer fields and community centers in Frogtown and Chinatown. Emphasis on preserving the wild—whether in faraway national parks or in Sierra Madre backyards—took them only so far politically. It scored them only few local successes and proved positively useless in attacking the overall idea and structure of flood-control engineering in the 1970s. But they have learned from their inner-city collaborators that the rivers—even if they cannot be restored to some pure natural state—can be made considerably better than they are now. And that is no small feat.

Thus, by 2002, FoLAR's forty-year art project to improve the river seemed well under way. Since the late 1990s, twenty-one new parks have been built along the Los Angeles River. In 2001, funding for river restoration topped a hundred million dollars, up from zero in 1985, the year of FoLAR's birth. Numerous organizations were busy working on creative solutions to a myriad of river issues. Northeast Trees, for example, hired at-risk youth to do construction work on eleven small parcels on the banks of the Los Angeles River; called pocket parks, they feature paths, benches, rock gardens, native plants, historical markers, yoga meditation guides, and sculpture by local artists, all of which contrasted with the chain-link fence, barbed wire, and KEEP OUT signs that had severed neighborhoods from the river since the 1940s.[109] Elsewhere, activists were putting the organizing model pioneered by the Chinatown Yard Alliance to work in other settings. A similarly diverse coalition of environmentalists, churches, civil rights groups, and community organizations formed to turn Taylor Yard into the second large parcel in the Greenway, and the newly formed Friends of the San Gabriel River opened a search for a sixty-thousand-dollar-a-year executive director, among whose qualifications were to be the ability to speak Spanish and experience in community organizing.[110] With all this activity, river activists were optimistic about the future of the rivers.

SEEDS IN THE CORNFIELD

One of the Chinatown Yard Alliance's lawyers hailed the Cornfield victory as "one of the most important things to happen in the history of L.A."[111] The attorney could perhaps be excused his hyperbole, for the Cornfield saga capped three decades of political and environmental turmoil that culminated in the region's most substantial departure from technocratic river management. The change began in the 1970s with environmentalists who sought to shift the flood-control discourse from technical grounds to aesthetic ones and who demanded a voice in decision making. The floods of 1978 and 1980 gave pause to the technical community and opened the engineers to reconsidering flood control. Finally, 1990s activism envisioned the rivers as resources helpful in meeting a variety of urban needs, from environmental preservation to neighborhood revitalization. At the end of these years, much had changed. In 1969, flood control was strictly an engineering problem. By 2002, it was a youth problem, a recreational problem, a housing problem, an education problem, a democracy problem, a racial problem, a quality-of-life problem, an aesthetic problem, and an environmental problem; in short, it was a social problem. In 1969, only engineers concerned themselves with flood control. By 2002, the engineers were joined by environmentalists, neighborhood activists, landscape architects, teachers, artists, poets, state park officials, bicyclists, and soccer clubs. Technocratic flood control was beginning to look a lot different.

The new shape that Los Angeles flood control appeared to be taking at the dawn of the twenty-first century was in many ways an outgrowth of broad social and political shifts that had occurred in postwar American society, as citizens reacted against the highly technical, closed-door, expert-driven decision-making style that characterized much policy making in the twentieth century. Southern California river activism, however, modeled an alternative to that style. Whether the Greenway and the public input to create it turn out to be substantially different from twentieth-century flood control—or indeed whether they materialize at all—remains to be seen. What is certain, however, is that during the 1990s, river activists defined an ambitious, yet perhaps achievable, dream: fixing Los Angeles by fixing the rivers. That may, in fact, be the culmination of MacAdams's forty-year art project. In the meantime, however, the river was changing southern Californians even as they sought to change it.

EPILOGUE

The Historical Structure of Disorder

Urban Ecology in Los Angeles and Beyond

Strange things have been happening in cities lately. In 1996, power lines sagging against tree branches outside Portland, Oregon, combined with other small power failures to trigger a cascade of blackouts that shut down law firms in Los Angeles, the airport in San Francisco, and casinos in Las Vegas. Before the lights went on again, four million people from Calgary to El Paso had lost power, some for up to sixteen hours. Meanwhile, East Coast residents have had their own utility troubles. Since February 2000, manhole covers have been spontaneously and inexplicably exploding from the streets in the wealthy and historic Washington, D.C., neighborhood of Georgetown. Some have been catapulted as high as thirty feet, flying like flat cannonballs into passing cars and store windows, while smoke and flames leap from the gaping holes in the street. Elsewhere on the East Coast, common mosquitoes, which have adapted to thrive in fouled waters that collect in rain gutters, old tires, swimming pool covers, bird baths, lawn furniture, and anywhere else that city water puddles, have since 1999 carried the deadly West Nile virus, which they pick up when feeding on infected crows, sparrows, and other urban wildlife. And last but perhaps strangest of all, in July 2001 a chemical-laden freight train burned out of control for days in a Baltimore tunnel while sixty million gallons of water gushed from a broken main into downtown streets and buildings. After the train derailed and caused the water main break—or was it the other way around? No one knows—utility officials wrung their hands for days, unable to stop the flood for fear that the necessary excavation would cause the burning tunnel to collapse onto firefighters. Making matters worse, the city had no plan for dealing with chemical spills in the tunnel, which is a key conduit for regional north-south rail traffic and telecommunication cables. But the city avoided a major environmental disaster, as water from the broken main and fire hoses diluted the acid that leaked from the railcars, sweeping it away from the flames and into the city's storm sewers and harbor. As it was, the flood and fire shut down central Baltimore for several days, while the effects rippled up and down the East Coast, and Internet service was disrupted as far away as Africa. A month later, manhole covers began blasting from Baltimore streets, apparently as a result of underground explosions of chemicals leaked by the train. In December 2002, as the National Transportation Safety Board wrapped up the fact-finding phase of its investigation, the causes of the water and fire disasters remained unknown. What is clear, however, is that their concurrence made the crisis more severe than it would have been had either occurred by itself.[1]

Apparently Los Angeles is not alone. It is not alone in watching small errors explode into big problems. It is not alone in depending on unrelated systems that unexpectedly interact. It is not alone in generating processes that both create and sustain urban life while simultaneously threatening it. It is not alone in suffering remote causes that trigger faraway effects. And it is not alone in creating technical systems that fail in strange and unforeseen ways, sometimes in conjunction with nontechnical ones. Although the particular circumstances that produce floods and other disasters in Los Angeles are of course unique, the city is by no means extraordinary or exceptional in catastrophes—the frequent claims for Los Angeles' exceptionalism notwithstanding. *First, worst, largest, last, most,* and other superlatives pervade both lay and scholarly descriptions of the city, especially when it comes to disasters. Certainly in popular conception, Los Angeles is more closely associated with catastrophe than any city in the world. Think earthquakes, think Los Angeles—despite the fact that it has never had a really devastating quake, the kind that kills thousands and razes entire cities, the kind that cities in Japan, the Middle East, and Central America experienced over and over in the twentieth century. Los Angeles quite possibly never will. In fact, despite its apocalyptic reputation, the city is actually quite ordinary in its disasters. Its droughts are not Sahelian; its floods would be puny on the Mississippi; and its earthquakes have not approached the magnitude, death toll, or devastation of several recent Asian temblors.

The lesson that Los Angeles flood control offers, then, is not a jeremiad about impending doom but rather an insight into the apparently random but actually patterned omnipresent threat of low-grade failures that afflict modern cities. Everywhere we look, it seems, we find disorder in urban places, disorder that challenges the assumption that lies behind all environmental management—that natural systems are more or less predictable and that the exceptions to this rule can be controlled by the rationality of human artifice. But disorder is not the same as pure randomness; it is not without structure. The significance of Los Angeles flood-control history is that it helps explain the historical structure of disorder in cities.

THE UBIQUITY OF DISORDER

Before we explore that structure further, however, it is useful to get a sense of just how widespread the kind of disorder is that Los Angeles flood control exhibits. One historical source of disorder that Los Angeles shares with other places is its demand for predictability from natural and technical systems that are heavily influenced by unpredictable political and social systems. Take San Antonio, for example. In 1946, ten people died and thousands more found themselves homeless after a flood swept through poor neighborhoods on the city's heavily Latino west side. Meanwhile, the rampaging water spared the downtown. This distribution of distress, however, was hardly an act of God or a random fluke of nature. Rather, the commercial center of the city was protected by Olmos Dam, completed in 1927, six years after a similar flood had damaged both the downtown and the west side. In fact, the grand buildings and the later-famed River Walk that occupied the city center in 1946 were possible at all only because of the flood protection provided by the dam. Lost in this flurry of protections in the 1920s and 1930s, however, were San Antonio's poor Mexican American residents, who watched as the city expended a few thousand dollars for flood control on the west-side creeks while it appropriated millions for Olmos Dam. Not until the late 1960s and early 1970s, when the U.S. Justice Department ordered reforms that insured greater west-side representation in city government, did effective grassroots organizations bring about flood protection for the west side.[2] As in Los Angeles, then—though the timing and particulars were different—normally unrelated social systems interacted. Where, when, and how water flowed in San Antonio had as much to do with ethnic relations and distribution of political power as it did with engineering, water, or weather.

Timing, though, is not just an insignificant detail. As a comparison of Denver and Los Angeles flood control shows, it can make all the difference. The Mile-High City confronted its flood menace for the first time in 1965, when an overflow of the South Platte River left twenty-six dead and $325 million in property damage. As was frequently the case in Los Angeles history, the tragedy spurred a technocratic response: the construction of an Army Corps dam upstream. In Denver, however, the flood also inspired the development of the Platte River Greenway, a string of paths, bikeways, landscaped parks, museums, amusement parks, and boating facilities along the river, which have emerged piece by piece in the decades since the floods and become an integral part of the downtown's economy and aesthetics.[3] In contrast to Los Angeles, where the first efforts at comprehensive flood control coincided with a time when technical experts enjoyed considerable public prestige and power, Denver confronted its river at the height of the environmental movement. Not surprisingly, the strategies the two cities initially pursued differed significantly. As it did in San Antonio and Los Angeles, then, the timing of flood-control efforts, and more specifically who happened to be in power at that moment, left lasting consequences in Denver for riparian land use and the flow of urban waters.

Apparently the influence of nontechnical concerns over the flow of water is not limited to the world of market capitalism, because politics also is the driving force behind flood control in China. Three Gorges Dam, which will rise 600 feet and span 1.3 miles across China's Yangtze River, is currently the largest construction site in the world. A 380-mile lake will eventually back up behind it, drowning nineteen cities and displacing nearly two million people; its generators will churn out fifty times more power than the world's next largest hydroelectric facility; by decreasing the flow of freshwater to the Pacific, it will likely warm the climate in Japan, thousands of miles away; and it will protect cities downstream on the Yangtze, along whose banks a quarter of a million people have died in floods over the last century. It's going to be massive. But the big dam does not lack critics. International engineers suggest it may be structurally and fiscally unsound. Environmentalists lament the vast tracts of habitat it will destroy. Human rights advocates champion the cause of the refugees it will dislocate. A string of smaller dams would likely be cheaper, less dangerous, and less devastating in terms of human and environmental impact while providing the same amount of electricity and flood protection. One thing the smaller dams would not offer, however, is a political statement. To the communist government that craves international respect and struggles to suppress the domestic dissidence that has been on the rise since the Tiananmen Square demonstrations in 1989, the dam is a monument to national greatness. It says to China's one billion people and to countries around the world that this government can build the world's largest dam, impose its will on the residents of the nation's interior, and tame the world's third longest river. Along with other megaprojects in the works—the world's highest railroad across the Tibetan Plateau, a twenty-five-hundred-mile natural gas pipeline, and the "Bionic Tower," which will house a hundred thousand people and dwarf the world's next tallest building—the dam demonstrates that China has joined the ranks of modern nations. Not unlike the Army Corps's promise to help save capitalism by putting people to work on flood-control projects in the 1930s, Three Gorges Dam has a political appeal to China's communist government that overshadows the technical, environmental, and social drawbacks.[4]

China's great dam has not yet even been built, much less burst, but catastrophes that result from volatile blends of technical and nontechnical failures are widespread. Government corruption allowed for shoddy construction of Mexico City apartment buildings, which collapsed with deadly results in a 1985 earthquake. Scores of low-income elderly residents died of heat exhaustion during Chicago's steamy summers in the 1990s. As temperatures reached one hundred degrees and humidity reached nearly the same percentage, senior citizens suffered alone in poorly ventilated un-air-conditioned public housing units behind doors and windows they had locked out of fear of crime, isolated from help by their poverty. They were victims not only of the rising mercury but also of society's inability to distribute the benefits of technology or even to mitigate the danger of opening a window.[5] Even Georgetown's mysterious manhole problems in part reflect social fears. The local power company upgraded underground utilities all over the city in the late twentieth century, taking advantage of the ripping up of the streets at that time for construction of Washington, D.C.'s Metro subway lines. Georgetown residents, however, successfully fought the extension of the Metro into their neighborhood in order to maintain their quarter's historic and aristocratic character and to keep the visiting riff-raff out; now they dodge exploding manhole covers instead.[6]

Finally, there is Los Alamos. Nestled in the mountains of northern New Mexico, Los Alamos was the victim of a National Park Service–

prescribed burn that escaped the control of its supervisors in May 2000. By the time the flames died down, more than forty thousand acres had burned, and 420 families were homeless. Actually, Los Alamos was an ironic place for a forest fire to burn a city, for in the not-too-distant past there was neither a forest nor a city there. In the late nineteenth century the top of Cerro Grande Mountain, the spot where the fire began, was a grassy meadow maintained by prairie fires that had historically swept the area and prevented the growth of more than a few scattered trees. The advent of sheep grazing, however, reduced both the grasses and the fires, and the forests quickly invaded. With the forests came fire suppression to protect the commercially valuable stands of ponderosa pine that replaced the meadow. In 1943, the federal government located its national nuclear laboratory in Los Alamos, and the town and its vicinity attracted a population of more than twenty thousand people by the end of the century. Those who could afford to flocked to new neighborhoods on the edge of the forest, attracted to the serenity of having wilderness in their own backyards. To the extent they thought about it at all, most who moved into the forest probably assumed it to be a primordial and static ecosystem, likely to endure in perpetuity. They never dreamed it was hardly stable or that their own presence made it less so.

So in the aftermath of the fire, the finger pointing started immediately, as desperate people sought a simple explanation and a target on which to pin blame for the billion-dollar tragedy. When the National Park Service board of inquiry issued its report a year later and found no broken rules or acts of negligence and recommended no individual punishment, almost everyone was displeased. "The Park Service should hold its personnel accountable," the *Denver Post* opined. "I find it hard to believe," fumed the Los Alamos congressional representative, "that no one is held accountable. Didn't someone make a mistake?"[7] Here was the corollary of what led the people to live in the forest in the first place: if nature is normally stable, then something really bad must have happened to produce so much distress. The possibility that it did not take a colossal blunder to produce an equally colossal disaster flew in the face of most people's understanding of nature and natural disasters.

But the explanation was far from simple. The fire was caused by a complex of unfortunate mistakes and coincidences that the Interior secretary Bruce Babbitt likened to a series of stones loosened from a mountainside: "Sometimes a rock is dislodged and nothing happens, but other times a rock is dislodged and it starts a cascading series of events . . . [until] you have a landslide at the bottom." The first rock to dislodge came in the form of bureaucratic urgency. By the 1990s, officials at neighboring Bandelier National Monument eyed the overgrown forest and accumulating downed branches and other ground fuel around Los Alamos and saw a time bomb that awaited only the slightest spark. They even developed a plan for cutting trees to thin the forest but delayed in circulating it in anticipation of environmentalist opposition. By the spring of 2000, things had become urgent. Searing drought portended a severe fire season, and the park superintendent, who favored the use of fire as a tool in ecological restoration, was about to retire—and who knew whether his successor would continue that practice?[8] Cerro Grande had to burn—soon.

Amid this sense of urgency, the fire planners made several small but fateful mistakes. First they inadvertently followed an outdated Fire Complexity Rating Worksheet (a protocol for planning and executing prescribed burns) that had been accidentally posted on the Park Service web site. In addition, they also underrated some of the variables in the complexity rating—easy enough to do given the imprecise and subjective judgments that ratings require. Coupled with the use of the wrong worksheet, however, the faulty rating judgments led to a serious misclassification of the prescribed burn as less risky than it actually was, which in turn led planners to underestimate the backup resources that would be needed if the fire got out of control. That error, plus a miscommunication between the park officials and the dispatch center in charge of providing backups, meant that once the fire got out of hand, additional personnel and supplies were slow in coming. Meanwhile, contrary to standard practice, the reports that park officials obtained from the National Weather Service did not contain the long-term wind forecasts that would have warned the fire supervisors of the pending gusts that eventually came up after they set the fire. Organizational urgency, incorrect web sites, small errors of judgment, miscommunications, and the remarkable coincidence that the incomplete NWS reports happened to come on a day when the Park Service was doing a prescribed burn and when the winds were about to change—these were the stuff that created very dangerous circumstances as officials ignited the burn.[9]

Even under the best of circumstances fire is unpredictable, and these were hardly favorable conditions. As winds came up, it appeared the fire would get out of control. Officials started a backfire to control it, and for a moment it looked like all was contained. Although it is not known exactly what happened next, some say a single ember—the smallest of causes imaginable—flashed in a gust of wind, and the conflagration that would chase Los Alamos residents from their homes was on. Despite the public cry for a scapegoat, there was no single negligent person to blame, no errant policy, not even extraordinary misbehavior by Mother Nature. Instead, according to the Park Service board of inquiry, the fire resulted from the "normal prescribed fire activities," any one of which "may not have been problematic" but which together "proved seriously inadequate."[10] Indeed, the cause was systemic, the disorderly result of a system thought to be orderly.

From San Antonio to Baltimore, Chicago to Shanghai, and Los Alamos to Los Angeles, modern cities face an array of environmental hazards, many of which stem at least partly from things that have nothing to do with water or weather. Poverty and social isolation sentence elderly Chicagoans to bake in their apartments. Cities along the Yangtze are sacrificed to the Chinese government's appetite for big engineering monuments. Falling tree branches near Portland shut the lights off in cities all over the West. San Antonio finds enough money to protect the prosperous downtown but not the low-income west side. When fire and flood coincide in Baltimore tunnels, the Internet goes down in Africa. And in Los Angeles, city building aggravates the flood hazard; fifty-year rains sometimes happen several times a decade; and flood-control devices fail before reaching their design capacity. If such disorder is as ubiquitous as it seems, then the world is perhaps put together quite differently from how we think it is.

THE HISTORICAL STRUCTURE OF DISORDER

But, then, how *is* it put together? One explanation, of course, is chance, but the proposition that the structure of nature and history contains some significant random elements makes many people nervous, and so we instead seek other explanations. Often we point the finger at nature. As the *Chicago Tribune* editorialized during the 1998 hurricane season, in the face of "nature's most violent displays of brute force," humans "can do little but watch in awe the 'great mischief' of Mother Nature." In other instances, such as the public demand for a hanging after the Los Alamos fire, we blame human misdeeds. In *Ecology of Fear,* for example, Mike Davis faulted the "selfish, profit-driven presentism" of civic leaders and the business community for "killing the Los Angeles River" and preventing the implementation of a flood-control plan that would have preserved the natural ecosystem. Sometimes we blame both. John McPhee's *Control of Nature* juxtaposed southern California's awesome and inexorable geologic and climatic forces with the puny and hubris-filled efforts of humanity to tame them.[11] Although disasters do seem to derive from a fair amount of chance, environmental unruliness, and human error, all of these explanations remove disasters from history; that is, they separate the events from the contexts in which they occur. This is a distortion. Greed, hubris, and the brute force of nature are transhistorical elements that are omnipresent but manifest themselves differently depending on historical contexts. The political arrangements of a given time and place, for example, empower certain people to act on their bad judgment, greed, or hubris more effectively than others; and economic and social organization helps to decide toward what ends human greed and hubris will be directed. If any one event, such as a flood, or any one timeless feature, such as the power of nature or human foibles, is selected for analysis by itself, it turns out to be connected to everything else—social relations, political structures, economic organization, environmental conditions, and all the little quirky events that happen under those circumstances. The best word in the English language to describe such a system, in which everything is connected to everything else and in which change in each element both responds to and generates changes in the other elements, is *ecological.*

The Historicity of Nature

The claim that human history is ecological is not the same as environmental determinism. To describe Los Angeles, or any city, as an urban ecosystem is not to relegate history to the outcome of some Darwinian struggle or to describe it as a march from a simple village society toward an urban climax community. Rather, in recent decades, scientists and environmental historians have worked out new understandings of nature, molding ecology into an increasingly historical science, rich in its consideration of the contingency and complexity of the objects it studies. In doing so, they have constructed new ways of finding order in the evolution of natural systems. These insights about nature have profound implications for how we understand the flow of water in Los Angeles and the society that has grown up around it.

First of all, nature is highly contingent. Even natural selection, long considered the grand force propelling life on earth toward ever more complex and highly adapted forms, it turns out, is highly sensitive to chance. Some biologists now suggest that the diversity of life has risen and fallen drastically throughout the earth's history and that individual species today result as much from accidents as from the inexorable forces of natural selection. A meteor strikes, suddenly changing the global climate and along with it the rules of natural selection. Adaptive traits become disadvantages; accidents of evolution suddenly become keys to survival under the new circumstances.[12] The evolutionary biologist Stephen Jay Gould had a favorite metaphor: the tape of life. Rewind it, erase everything that happened, play it again, and it will never come out the same way twice.[13] Change one little thing and everything else will turn out differently. No law of nature guaranteed the evolution of mammals or the demise of the dinosaurs. No law guaranteed the New Year's Eve storm either. Go back in time, redirect the wind slightly or change the temperature a little, and the storm will ascend the mountain slopes. Residents in the foothills will sleep safely and soundly through the night. From the fate of the dinosaurs to the foothill floods, natural events are highly sensitive to even slight changes in conditions.

This sensitivity exists because nature is also so complexly organized. The discipline of ecology once had its own theory of evolution: succession. Ecological communities, according to this paradigm, start out as disorganized, unstable groupings of individual plants and animals. Gradually over time, they succeed through fixed stages toward "climax communities." Such communities have characteristic vegetation types that are homogeneous throughout their ranges and are uniquely in balance with the climate. After a disturbance, such as fire, the community proceeds back toward its climax conditions. Succession, in the eyes of ecologists, however, has grown considerably more complex. "Prior to the 1950s," the plant ecologist Michael Barbour observed, "nature was simplistic and deterministic; after the 1950s, nature became complex, fuzzy edged, and probabilistic." Ecologists now see ecosystems not as discrete homogeneous units but rather as overlapping ranges of species, each with its own reproduction, competition, and survival patterns combining to form a unique collection of interacting individuals. Nature is not so much interdependent—changing as a whole according to some progressive pattern—as it is interactive—with lots of different components changing according to their own logic as well as in response to the independent changes of other components. If you want to understand the forest, you have to understand the trees—and the soil, and the animals, and the water, and the microclimates, not to mention all the other ecosystems it interacts with. If you want to understand the river, you have to understand the vegetation, channels, weather, and soil, all of which are changing in different directions, at different speeds, and according to different logics. The dizzying number of variables, forces, and counterforces makes the behavior of even a small space of nature, even the flow of a single river, appear hopelessly random.[14]

But it is far from random. As Gould pointed out, any outcome E *has* to follow from antecedent events A through D.[15] There is order in nature; it's just that that order is very different from what we usually imagine. Newton revolutionized science when he suggested that, given knowledge of the right parameters, scientists can predict the paths of objects in motion. Since then, the discipline of physics has flourished by extending the principle that events can be understood and predicted on the basis of knowledge of previous conditions. The hundred-year flood concept and other flood-control knowledge were predicated on similar premises. Even physics, however, has had its holes. "There are parts of science that are very exact," said one flood scientist. "Predicting eclipses 10 years in advance works that way. Predicting how rivers behave doesn't work that way."[16] Physics could describe collisions of billiard balls, and even subatomic particles in an accelerator, but had trouble modeling the turbulent flow of water. And even with mountains of data, the laws of physics did not predict the behavior of weather systems for more than the shortest of periods.

Beginning in the 1950s and accelerating since the 1970s, study of such mysteries has yielded astonishing results.[17] A meteorologist simulating weather on a computer found that changing numerical input parameters by even the tiniest fraction yielded wildly different outcomes. Scientists accelerated and decelerated pendulums with constant and regular forces only to find that the pendulums moved in patterns for only short periods of time before shifting into unrepeated sequences. Loosely grouped under the heading "chaos theory," such findings contradicted conventional scientific wisdom that had held that small causes do not blow up into large effects. The research also challenged the mathematical version of the climax community—the principle that systems with regular forces acting on them eventually settle into regular behavior. Traditional physics explained much of the world, but some corners of nature are messy and nonlinear.

Southern California is one of those corners. Whenever the flows of rivers or flood-control policies suddenly change direction, whenever well-designed debris basins work like flip buckets, whenever storm cells drench single canyons with thousand-year rains, whenever historical conditions conspire to render some flood-control strategies appealing and others invisible, the interaction of nature and society resembles the behavior of species populations, weather systems, mass extinctions, and pendulums. History works ecologically.

The physical and cultural elements of the city add another layer of complex organization and contingency to nature's. In *Normal Accidents,* the social scientist Charles Perrow investigated the organization of highly technical systems like nuclear power production. He found that such systems are complexly organized; that is, they contain large numbers of component parts that can interact in an incalculable number of ways, linking parts of the system that were not previously connected. For example, a nuclear plant's cooling system malfunctions at the same time as the alarms fail, and meltdown results. In more everyday terms, you lock yourself out of your car on the day of the bus strike, so you miss an important job interview. In both cases separate, minor, but coincidental failures in a subsystem and its backup multiply mishap into catastrophe. Such systems, Perrow found, also are tightly coupled; that is, every step in a given operation depends on successful completion of each previous step. One failure quickly generates another. A galley fire causes multiple shorts in an airplane's electrical system, which in turn short out the fire alarm.[18] In such tightly coupled systems, there is only one way for things to happen successfully. At the same time, however, complex organization multiplies the number of ways things can go wrong. Together the two characteristics are an explosive mix.

The Los Angeles urban ecosystem is both complexly organized and tightly coupled. Before the city rose on the coastal plain, shallow, shifting riverbeds periodically overflowed and carved new channels, harmlessly distributing their waters and fertile silt across the plain. Water could reach the sea by any number of different ways—or even not reach it—and the system still functioned just fine. The urban ecosystem's high-leveed cement channels, however, are more tightly coupled. Water must stay within them, or else overtopping will undermine the levees, precipitating disaster. Law was just as tightly coupled as engineering. The Flood Control Act of 1915 required that county engineers build whatever voters approved by referendum; consequently, when citizens began to struggle over the design and location of San Gabriel Dam, *after* voting to approve the project, there was no way to resolve the dispute. Five years of political stalemate delayed construction. Both flood-control law and engineering structures design nature to work in a particular way. When it does not, failure occurs.

The urban ecosystem is also complexly organized. Levees collapsed in the 1980 flood in part because they were not constructed according to design. Redesigning them entailed traversing mountains of government bureaucracy in order to reauthorize the project according to different specifications. It was easier to build substandard structures and hope that the waters would not test them. The storms rolling off the Pacific, however, linked the flood-prone climate to the inflexible bureaucracy. Water rose and ate away the banks, and the levees crumbled.[19] In the flood, unrelated environmental and government structures suddenly interacted, with significant consequences.

Thus, the Los Angeles urban ecosystem is not merely a better-designed version of messy nature. Instead, complexity increases the number of ways that water can flow destructively, and at the same time, tight coupling requires that it flow exactly one way. As a result, the system is prone to failures, especially floods.

Patterned Disorder

Complexity and contingency in the Los Angeles urban ecosystem stem from patterned interactions among historical events and structures. The structures—for example, political arrangements, booster culture, climate, or even human greed and hubris—both generate and are generated by particular historical events. Floods and other displays of the awesome power of nature are among these events, but so are municipal elections that provide funding for flood-control programs, founding of environmental organizations to lobby against existing flood-control practices, or shifting of flood-control responsibility from local to federal governments. Structures enable certain events and constrain others, and events alter some structures while reinforcing others. Thus, over time, certain events occur, and flood-control institutions and other structures of society both persist and evolve.[20] These interactions constitute historical change. So in order to understand the urban ecosystem—how it emerged, how it works, how it has changed—we need know how the structures and events fit together, how they interact to produce vulnerability to floods and to shape humans' responses.

There were at least three patterns in the urban ecosystem whereby ordinary events and structures combined to generate vulnerability and response to catastrophe. The first was feedback: the very features of the system that created and sustained the city also made the city vulnerable to devastation. The second was intricacy: the normally separate components interacted in numerous quirky ways to spread and escalate vulnerability. And the third was history: the historical conditions under which those interactions occurred constantly changed, so that an event that was harmless in one context sometimes could become hazardous in another. Together, feedback loops, intricacy, and historical dynamism outline the structure of disorder in the Los Angeles urban ecosystem.

As an example of feedback, consider the fire that denuded the hillsides the month before the storm of New Year's Eve 1933–1934. For centuries, the ecology of the foothills had made the slopes fire-prone. The blazes actually spurred the growth of chaparral, a particularly fire-prone plant community. The chaparral produced more fire. The fire produced more chaparral. Thus, the climatic and ecological structure of foothill chaparral caused fires, which in turned reinforced the structure.

Urban development worked similarly. First, in the late nineteenth and early twentieth centuries, the very process of taking advantage of nature's hydrologic blessings to build the city speeded the end of those blessings and increased the flood threat. Rapid immigration, land development, and dispossession of the Mexican population—the very actions that built the city—combined to sustain and aggravate the flood threat. Later, in the 1920s and 1930s, the very mountains that produced the floods also made the foothills attractive for settlement, which both increased the flood potential and increased the need for flood protection. Moreover, the rugged and unstable mountain geology also made big flood-control dams technically infeasible. Finally, throughout the century, flood control enabled development, which aggravated the flood problem, which in turn required more flood control. And the cycle went on. Immigrants, subdivisions, scenic views, and flood-control works were commonplace in Los Angeles. They were not external or extraordinary parts of the urban ecosystem but rather were integral to its creation and continuing existence. Their interactions with one another and with structures such as ethnic prejudice, a pro-development ethos, and a cloudburst-prone climate, however, escalated vulnerability to floods and decreased society's ability to protect itself.

Intricacy in the relationships among subsystems also contributed to the urban ecosystem's vulnerability. In the early 1920s, for example, drought and urban development linked climate and economy to produce a sudden, and ultimately infeasible, shift in flood-control policy, leading the Flood Control District to attempt to build the San Gabriel Dam. The controversy that the project sparked, in turn, spread into other segments of the flood-control program. Ostensibly, the 1926 ballot measure for additional flood-control bonds to finance the protective works for the La Cañada Valley had nothing to do with the San Gabriel Dam, but the election coincided with the rising unpopularity of the dam, and thus the voters defeated those new projects. The fire and rainstorm of New Year's Eve struck the La Cañada Valley before the flood-control program could get itself straightened out. Finally, once the flood reopened the question of how to control the rivers, the Great Depression–era enthusiasm for big public works projects and the political structure of Los Angeles combined to ensure that technocratic solutions would continue to dominate flood-control policy. In each case, coincidental timing of otherwise unrelated events suddenly linked separate parts of the urban ecosystem, so that small flaws in one spread into another, escalating along the way.

Finally, the urban ecosystem's historical dynamism added to its vulnerability. Political and technical conditions that were not problematic (and in fact were even advantageous) in some contexts escalated into big problems when conditions changed. The 1915 Flood Control Act, for example, prohibited changes to engineering plans once they were approved. This prohibition was an effective, and perhaps even necessary, measure to generate support for flood control among Progressive Era voters and politicians who worshipped independent commissions of experts and who mistrusted special interests that might interfere with the experts' objectivity. The same prohibition, however, served to thwart flood control later on, once disagreement among the experts stalled the San Gabriel Dam and discredited the entire flood-control program. Similarly, the closed-door, hierarchical, expert policy making that was so effective in turning blueprints into flood-control devices during the politically quiescent midcentury decades grew unpopular after the rise of environmentalism. Physical structures worked similarly to the political structures: the check dams above La Cañada Valley, the San Jacinto levees, and the foothill debris basins successfully controlled debris and water under conditions of moderate rainfall. Under more extreme circumstances, however, they not only failed to control floods but in some cases actually aggravated them. Thus, in both politics and engineering, strategies that advanced the flood-control cause in some historical contexts increased the hazard in others.

Thus, both floods and flood control had roots in the historical interaction between the specific events of the flood-control process and the structures of nature and society. The floods were ordered events that grew out of feedback loops, intricate relationships, and historical dynamism of the interactions between nature and the choices people made about how to arrange society. Similarly, the public political will for flood control, the types of strategies that emerged, and the ability of society to imagine and execute those solutions depended not merely on technical factors but also on the timing of floods, the cultural context, and the political terrain. None of this changed with the paving of the rivers. Feedback loops still amplify small causes into big effects. Apparently unrelated components of the urban ecosystem still interact. And because of changing historical conditions, the pursuit of effective flood control sometimes resembles shooting at a moving target—what works under one set of conditions may not under others. These patterns shaped not only the frequency and severity of floods but also the range of solutions that southern Californians could imagine and implement. Water still floods in Los Angeles, then, because feedback loops and intricate and historical relationships among events and structures have combined to make bulldozers and concrete both appealing and ineffective.

WHAT LIES DOWNSTREAM

All this complexity and contingency leaves one to wonder, though, if the hazards afflicting urban ecosystems such as Los Angeles are simply too complex and overwhelming to be managed effectively. Perhaps so, but in some ways we have no choice. We have to manage them. After the massive worldwide urbanization that took place during the twentieth century, more and more human beings are living in these kinds of circumstances. Moreover, every indication is that disasters are on the rise, despite the remarkable advances in prevention and recovery technology achieved during the same period. The number of annual natural disasters has steadily increased since the 1970s, as has their total annual cost (in fixed dollars). The worldwide price tag for weather-related catastrophes in 1998 alone topped that of the entire decade of the 1980s.[21] To make urban living in southern California and elsewhere possible at all, humans depend on flood control and other complex technical systems that in turn depend on equally complex political, social, cultural, and economic arrangements. Such systems provide very basic things such as food, shelter, sanitation, recreation, transportation, communication, and safety. They also, however, sometimes fail, and even when functioning properly, they often aggravate social divisions and degrade the environment. Thus they simultaneously enhance and impair urban life, and the urban ecosystem in Los Angeles in the twentieth century featured plenty of both. The challenge remaining is to improve such systems to create a better chance that in the future they will enhance more and impair less.

In that endeavor engineering will surely continue to play an important role. Flood-control devices have saved many lives in southern California and prevented billions of dollars of property destruction. Without some sort of flood control, urban life in Los Angeles and many other cities could not have developed and could not continue. In fact the central reason that natural disasters in Los Angeles and the rest of the United States have been so much less devastating than in many other countries is the heavy American investment in hazards engineering. Flood devastation in Central America and Southeast Asia offers a frightening picture of what a world without adequate hazards engineering would look like. The solution to the flood-control problem, then, is not to fire the engineers, rip up the concrete, and let the rivers do whatever they will.

Nor, however, is the solution continued technocracy. Institutional arrangements that define problems strictly in technical terms and that empower only people with technical expertise to deal with them are structurally unsuited to cope with patterned disorder. Feedback loops and intricacy make both the technical and nontechnical aspects of flood control highly volatile, but technocratic responses have been designed to ignore or suppress that volatility. And while historical dynamism constantly shifts the environmental and political ground rules of flood control, the numerous comprehensive plans launched since 1915 have sought once-and-for-all solutions. The urban ecosystem is flexible, and technocracy is rigid. Not surprisingly, assembly-line flood control has had only partial success.[22] Even when it succeeds on its own terms, technocratic flood control carries a high price tag: overreliance on engineering devices that sometimes fail; empowerment of people with technical expertise and exclusion of people with historical, ethnic, environmental, educational, and other sorts of contributions to make; prevention of community involvement in river management and unequal access among groups to the levers of policy making; aesthetic and ecological scars through the heart of the city; and isolation of low-income neighborhoods physically and politically, often in the name of benefits enjoyed disproportionately by wealthier citizens.[23] Despite the many benefits technocratic flood control has secured, these costs are too high.

The solution then, is to make flood control more ecological. Here, the Chinatown Yard Alliance provides a promising model. The activists who managed to combine flood control, environmentalism, and urban regeneration at the Cornfield imagined the river as being embedded in complex environmental and human landscapes. They measured costs and benefits not by merely comparing the price per cubic yard of concrete to the value of urban development to be protected but rather by envisioning the river's possibilities for improving the community. They saw in the river not simply flow rates, water volumes, and levee dimensions but also recreation, education, pollution abatement, historical memory, beauty, and justice. Technocracy failed to incorporate those things into river management; furthermore, it also often dismissed them as irrelevant, subjective, and naive—unworthy of the attention of technical experts who had serious work to do. The engineers can hardly be blamed though. It would have been daunting, to say the least, for them to integrate their technical program with land-use regulation, insurance practices, community empowerment, parks and recreation enhancement, environmental protection, public education, historic preservation, and the full range of urban life to which the rivers were connected. Moreover, even if engineers had been inclined to do so (and few were), they were both untrained and unauthorized to undertake such a wide-ranging program. No one—not engineers or environmentalists or community organizers—can, by themselves, understand the whole river and all of its connections to the city. Urban ecosystems are simply too complex.

The secret to the Chinatown Yard Alliance's broader vision, then, is political. The alliance found a way to include in the debate over the Cornfield many voices, each of which spoke to a different aspect of the river's place in the urban ecosystem and which collectively came from antagonistic viewpoints to find common ground in Chinatown. Engineers brought technical expertise but had to broaden their imagination of rivers as simply machines. Environmentalists brought political connections but had to abandon their fantasy of restoring pristine nature. And neighborhood organizers brought legitimacy of speaking for the people but had to overcome their suspicion that the other groups were out to get them. In contrast, technocratic flood control has required centralized hierarchical political structures in which experts make decisions with minimal public involvement, conditions such as those that prevailed in the midcentury decades. Insulated from political conflict, midcentury engineers were not forced to consider competing visions, and consequently they did not confront the political or technical limitations of their endeavors until the 1980s. Without conflict, no new river visions could come up for consideration. At the Cornfield, however, after nearly a century of unflagging commitment to assembly-line flood control, a cacophony of dissonant voices forced their way into the policy debate, and something unprecedented happened.

The Cornfield, however, is but one plot of land at one spot along one river. What kind of management will emerge at other places along the Los Angeles River or along the San Gabriel or the Arroyo Seco or the hundreds of other waterways in the region? What will happen to places that lack such large parcels of vacant riparian land? What will happen in cases where the interests of engineers, environmentalists, and urbanites do not coincide so neatly? The small amount of flood control the Cornfield will provide will supplement existing strategies, but what will happen when environmentalists demand that river greening substitute for engineering? And what will happen when the weather changes? The flourish of alternative river visions after 1998 coincided with dry years. Will there still be political will for nontechnical solutions when catastrophic storms return, or will Los Angeles then revert to solving every river problem with bulldozers and concrete? The flood-control road does not end at the Cornfield. An ecological approach to flood control—if one emerges at all—will provide no once-and-for-all solutions. Rather, it will be an ongoing process of imagining and reimagining the rivers and the city and of negotiating and renegotiating their management. Only by recognizing rivers as embedded in larger human and environmental landscapes and by incorporating many voices into the policy-making process will flood controllers be able to manage the variability of patterned disorder in the urban ecosystem.

In this regard, Los Angeles flood-control history offers not only a signal of things gone wrong, as the city's critics so often point out, but also a glimpse of things going right. The ecological and social justice ethic emerging among river activists in southern California at the beginning of the twenty-first century and the variety of public and private institutions put in place to carry out new visions for the river have the potential to reinvent not only the rivers but also environmentalism, hazards management, and perhaps all of Los Angeles along more humane and environmentally workable lines. By doing so, they offer an alternative form of institutionalization of flood control, a new idea and a new structure that differ from anything tried since 1914. The Cornfield site offers hope. Nothing has been accomplished there yet, but there is hope.

Notes

ABBREVIATIONS

CDPW:	California Department of Public Works.
CDW:	Charles Dwight Willard Papers, Huntington Library, San Marino, California.
CHR:	California History Room, California State Library, Sacramento, California.
CHSM:	City Hall, Sierra Madre, California.
DSCUCLA:	Department of Special Collections, Charles E. Young Research Library, University of California, Los Angeles, Los Angeles, California.
ENR:	*Engineering News-Record.*
FoLAR:	Friends of the Los Angeles River.
GPS:	Government Publications Section, California State Library, Sacramento, California.
HBL:	Henry Baker Lynch Papers, Department of Special Collections, Charles E. Young Research Library, University of California, Los Angeles, Los Angeles, California.
HL:	Huntington Library, San Marino, California.
HWRS:	Harriet Williams Russell Strong Papers, Huntington Library, San Marino, California.

JAF: John Anson Ford Collection, Huntington Library, San Marino, California.

JBL: J. B. Lippincott Collection, Water Resources Center Archives, University of California, Berkeley, Berkeley, California.

JDR: J. David Rogers Private Historical Collections, Pleasant Hill, California.

JPO: Manuscripts in the possession of Jared Orsi.

LAACC: Los Angeles Area Chambers of Commerce Collection, Regional History Center, University of Southern California, Los Angeles, California.

LACA: Los Angeles City Archives, Los Angeles, California.

LACBS: Los Angeles County Board of Supervisors, Executive Office, Los Angeles, California.

LACDPW: Los Angeles County Department of Public Works.

LACDPWTL: Los Angeles County Department of Public Works Technical Library, Alhambra, California.

LACFCA: Los Angeles County Flood Control Association.

LACFCD: Los Angeles County Flood Control District.

LAD: U.S. Army Corps of Engineers, Los Angeles District.

LAE: *Los Angeles Examiner.*

LAPL: Los Angeles Public Library, Los Angeles, California.

LAT: *Los Angeles Times.*

NAPR: National Archives, Pacific Region, Laguna Niguel, California.

NRC: National Research Council.

PAO: Public Affairs Office, Los Angeles District, U.S. Army Corps of Engineers, Los Angeles, California.

RNL: Richard M. Nixon Library, Yorba Linda, California.

TBE: Thomas Balch Elliott Papers, Huntington Library, San Marino, California.

TPL: Theodore Parker Lukens Papers, Huntington Library, San Marino, California.

USGS: United States Geological Survey.

WRCA: Water Resources Center Archives, University of California, Berkeley, Berkeley, California.

PROLOGUE. WATER IN LOS ANGELES

1. National Research Council, Committee on Natural Disasters, and Environmental Quality Laboratory, California Institute of Technology, *Storms,*

Floods, and Debris Flows in Southern California and Arizona, 1978 and 1980: Overview and Summary of a Symposium, September 17–18, 1980 (Washington, D.C.: National Academy Press, 1982), 11. There is some dispute about how the debris got there. Flood-control officials maintained that the water nearly overflowed the levees, leaving the debris behind as evidence. One Long Beach resident, however, claimed that waves well below the levees could have tossed it there; Friends of the Los Angeles River, "Proposed Flood Control Strategy for the Los Angeles and San Gabriel River Systems" [1995], unpaginated appendix, manuscript, LAPL; D. J. Waldie, "The Myth of the L.A. River," *Buzz,* April 1996, 84; George Stein, "Study Widens Risk Zone of Hypothetical Major Flood," *LAT,* 15 September 1987, sec. II, 1, 6.

2. This is a study about metropolitan Los Angeles. Given the amorphousness of that entity, a word on meaning is in order. For the most part, this study covers the coastal plain and mountain areas within Los Angeles County (excluding the Channel Islands and the high desert north of the San Gabriel Mountains). This spatial scope coincides with the jurisdiction of the Los Angeles County Flood Control District, the local government agency charged with flood control during the period covered by this study. In the pages that follow, therefore, the names "Los Angeles" and "southern California" are used more or less interchangeably to refer to the contiguously urbanized area within Los Angeles County. The people who live in this area I refer to as "southern Californians." I use "city of Los Angeles" to refer to the municipality of Los Angeles, and I use "Los Angeles Basin" to refer to the drainage area of the Los Angeles and San Gabriel rivers. Although urban southern California is centered in the city of Los Angeles, the contiguously urbanized space has expanded during the period covered by this study, as development has spread beyond the boundaries of Los Angeles County. Consequently, although the study does not systematically examine the adjacent counties of Ventura, Orange, Riverside, or San Bernardino, it does occasionally range into those places, especially in covering the recent period, when portions of those counties have become part of the contiguous urban space and thus effectively part of metropolitan Los Angeles. Consequently, in chapter 6 the definitions of "Los Angeles" and "southern California" expand slightly to include the areas outside the boundaries of Los Angeles County that have been incorporated by this metropolitan growth.

3. Theodore Osmundson, "How to Control the Flood Controllers," *Cry California,* summer 1970, 33.

4. Two hundred seventy million 1980 dollars are roughly the equivalent of 591 million 2002 dollars. Source: http://www.cjr.org/resources/inflater.asp, accessed on 28 February 2003.

5. "Storm Spawned by Shift in Weather Thousands of Miles Away," *LAT,* 11 February 1978, sec. I, 2, 26.

6. According to National Weather Service records, rainfall at the Los Angeles Civic Center between 1878 and 1992 was within about 15 percent of the annual mean only 25 times (22 percent of the time). Tree ring indices from the San Gabriel Mountains from 1599 to 1966 show rainfall being within 15 percent of the annual mean 114 times (31 percent) and within 10 percent of the mean 78 times (21 percent). Spanish mission crop data suggest that between

1769 and 1835, southern California rainfall fell within 15 percent of the mean 21 percent of the time and within 10 percent of the mean 12 percent of the time. According to Mike Davis, the annual precipitation in Los Angeles falls within 25 percent of the mean only 17 percent of the time. Andrew Rolle, *Los Angeles: From Pueblo to City of the Future* (Sacramento: MTL, Inc., 1995), 153; Linda G. Drew, ed., *Tree-Ring Chronologies of Western America: California and Nevada,* vol. 3 (Tucson: Laboratory of Tree-Ring Research, University of Arizona, 1972), 16; Lester B. Rowntree, "A Crop-Based Rainfall Chronology for Pre-instrumental Record Southern California," *Climate Change* 7 (1985): 327–41; Mike Davis, *Ecology of Fear: Los Angeles and the Imagination of Disaster* (New York: Metropolitan Books, 1998), 16.

7. It is interesting to compare the large number of assumptions that go into such calculations with the smooth statistical abstraction that the public sees afterward. The fifty-year flood is a neat and deceptively clear concept that can easily be represented on maps with firm lines. See, for example, the U.S. Army Corps's 1987 map of southern California with zones shaded differently to delineate the boundaries of the fifty- and hundred-year floods. The map promised two to four feet of inundation in some places, eight to ten in others; LAD, "LACDA [Los Angeles County Drainage Area] Update," September 1987, 6–8, PAO. Such exact representations hide the imprecision that goes into making these estimations of frequency. Consider, for comparison, a circa 1940 U.S. Department of Agriculture document on the frequency of storms in Los Angeles. At one point the author made an assumption for which he admitted "there is no scientific basis" but which he accepted anyway because it produced a curve that "looks reasonable." LACFCD, "Conditions at District Dams Following Flood of March 2, 1938," by Finley B. Laverty and N. B. Hodgkinson, March 21, 1938; see also appendix B, "Storm Frequency and Potential Storm Damage," in Department of Agriculture, Field Flood Control Coordinating Committee No. 18, "Volume I of Appendices: Appendices A to M to Accompany Survey Report of Los Angeles River" [ca. 1940], manuscript, box 239, RG 77, NAPR.

8. Tom Knudson and Nancy Vogel, "Superfloods Threaten West," *Sacramento Bee,* 23 November 1997.

9. Gerard Shuirman and James E. Slosson, *Forensic Engineering: Environmental Case Histories for Civil Engineers and Geologists* (San Diego: Academic Press, 1992), 189; J. Daniel Davis, "Rare and Unusual Postfire Flood Events Experienced in Los Angeles County during 1978 and 1980," in *Storms, Floods, and Debris Flows in Southern California and Arizona, 1978 and 1980: Proceedings of a Symposium, September 17–18, 1980,* sponsored by National Research Council, Committee on Natural Disasters, and Environmental Quality Laboratory, California Institute of Technology (Washington, D.C.: National Academy Press, 1982); Charles Pyke, "Return Periods of 1977–80 Precipitation in Southern California and Arizona," in *Storms, Floods, and Debris Flows . . . : Proceedings of a Symposium,* sponsored by NRC.

10. Department of the Interior, U.S. Geological Survey, *Flood in La Cañada Valley, California, January 1, 1934,* by Harold C. Troxell and John Q. Peterson, Water Supply Paper, U.S. Geological Survey, 796-C (Washington, D.C.: United States Government Printing Office, 1937), 54–55, 60–61.

11. A. E. Sedgwick, "Geology of the Forks Dam Site San Gabriel Canyon," March 1930, 3, manuscript, LACDPWTL.

12. Harold E. Hedger, "Harold E. Hedger Oral History Transcript: Flood Control Engineer," interview by Daniel Simms, 17, 19 May and 8 July 1965, pp. 20–21 (all quotations from this transcript © The Regents of the University of California and reprinted by permission of the UCLA Oral History Program, Department of Special Collections, Charles E. Young Research Library); John McPhee, *The Control of Nature* (New York: Farrar, Straus, Giroux, 1989).

13. John M. Tettemer, "Closing Comments on Debris and Sediment," in *Storms, Floods, and Debris Flows . . . : Proceedings of a Symposium,* sponsored by NRC, 484.

14. "Research Los Angeles County Flood Control," interviews with Los Angeles County residents, compiled by James W. Reagan, 1914–1915, 42, 48–49, 84, 142, 358–60, 513, manuscript, LACDPWTL. This is a collection of several hundred pages of interviews with old-timers about their memories of nineteenth-century floods in Los Angeles. The interviews were conducted by James W. Reagan, an engineer working on behalf of the county, and by his several assistants. See also James Miller Guinn, "Exceptional Years: A History of California Flood and Drought," *Publications of the Historical Society of Southern California* 1 (1890): 33–34.

15. *DeBaker v. Southern California Railway Company,* 39 *Pac. Rep.* 615 (1895).

16. LAD, "Los Angeles County Drainage Area Review: Final Feasibility Study Interim Report and Environmental Impact Statement," December 1991, revised June 1992, 40–41.

17. Jack Smith, "Driving Down the River," *Westways,* July 1974, 67.

18. Although popular understanding of ecology remains wedded to the deterministic paradigm of ecological succession proceeding toward monoclimax communities, first advanced by Frederick Clements in the early part of the twentieth century, the discipline of ecology has since the mid-1970s confronted the chaos and indeterminacy that influence many ecological phenomena. See Robert M. May, "Biological Populations with Nonoverlapping Generations: Stable Points, Stable Cycles, and Chaos," *Science* 186 (15 November 1974): 645–47; Daniel Simberloff, "A Succession of Paradigms in Ecology: Essentialism to Materialism and Probabilism," in *The Philosophy of Ecology: From Science to Synthesis,* ed. David R. Keller and Frank B. Golley, 71–80 (Athens, Ga.: University of Georgia Press, 2000); Frank Benjamin Golley, *A History of the Ecosystem Concept in Ecology: More Than the Sum of the Parts* (New Haven: Yale University Press, 1993); and Michael G. Barbour, "Ecological Fragmentation in the Fifties," in *Uncommon Ground: Toward Reinventing Nature,* ed. William Cronon (New York: W. W. Norton, 1995), 233–55.

19. On the relationship between politics, economics, and natural disasters worldwide, see J. M. Albala-Bertrand, *Political Economy of Large Natural Disasters, with Special Reference to Developing Countries* (Oxford: Clarendon Press, 1993).

CHAPTER 1. CITY OF A THOUSAND RIVERS

The source of the chapter title is the newspaper article "Unprecedented Rain Paralyzes Traffic; 1 Dead; City Streets Are Rivers, Hundreds Marooned," *LAE*, 19 February 1914, sec. I, 1–2.

1. *LAE*, 19 February 1884, in "Research Los Angeles County Flood Control," interviews with Los Angeles County residents, compiled by James W. Reagan, 1914–1915, 95, manuscript, LACDPWTL.

2. Ibid., 1a, 27, 73, 82, 84, 88, 94, 323–27, 387, 410.

3. Ibid., 54, 323, 526; "Los Angeles River Long Vexes City with Tricks," *LAT*, 11 May 1924.

4. "Research Los Angeles County Flood Control," 1a, 88, 323–27, LACDPWTL; "56,309,600 Tons Water Falls upon Los Angeles," *LAE*, 1 March 1914, sec. IV, 1; "Unprecedented Rain Paralyzes Traffic; 1 Dead; City Streets Are Rivers; Hundreds Marooned," *LAE*, 19 February 1914, sec. I, 1.

5. Although there are no definitive estimates for exactly how much property damage the earlier flood did, most estimates placed the figure around a few hundred thousand dollars. The estimates for the 1914 flood range between ten million and twenty million dollars, depending on the area taken into consideration and whether losses such as diminished property values and business costs due to interrupted commerce are included. To put the loss in turn-of-the-century context, the total cost of constructing the Los Angeles Harbor had been between $10 million and $15 million as of 1914, roughly $178 million to $268 million in 2002 dollars; LACDPW, "Los Angeles River Master Plan," June 1996, 311, LAPL. The geographer Blake Gumprecht offered a similar estimate in 1999, pegging the 1884 deluge damage at somewhere between 2.4 million and 16.3 million in 1999 dollars and the 1914 destruction in the neighborhood at $162 million (and perhaps double that in lost property values); Blake Gumprecht, *The Los Angeles River: Its Life, Death, and Possible Re-birth* (Baltimore: Johns Hopkins University Press, 1999), 157, 168.

6. Old-timers interviewed by Reagan almost unanimously agreed that the flood of 1884 (as well as those of 1862 and 1889) exceeded that of 1914 in terms of the extent of land inundated and the total volume of water discharged. See "Research Los Angeles County Flood Control," passim, LACDPWTL; Los Angeles County Board of Engineers Flood Control, *Reports of the Board of Engineers Flood Control to the Board of Supervisors Los Angeles County California* (Los Angeles, 1915), 4, Rare Books Collection, HL; hereafter all Los Angeles County Board of Engineers Flood Control sources are cited with "Board of Engineers." Engineers calculated the maximum discharge of the floods of the 1880s (the only ones for which they had measures) as 60 to 70 percent greater than that of 1914, thus confirming the old-timers' recollections. Engineers' estimates of the peak discharge of the San Gabriel River during one of the 1889 floods, for example, ranged from 35,000 to 47,000 cubic feet per second; calculations of the maximum discharge on the river in 1914 were fixed at only 26,680 cubic feet per second. On the basis of these and other calculations, engineers reported to the Board of Supervisors in 1915 that the ratio between the 1884 flood and that of 1914 was 1.65 to 1; the 1889 to 1914 ratio they calculated to be 1.77 to 1;

ibid., 4, 9, 145, 178. One engineer estimated the extent of the land that the 1914 flood overflowed to be only one-fourth to one-third that of 1884 and 1889; [James W. Reagan] to Los Angeles County Board of Supervisors, Los Angeles, 11 June 1915, Old Document Files 4970F, LACBS.

7. The source of the letterhead is Frank Wiggins to Los Angeles County Board of Supervisors, Los Angeles, 9 September 1915, Old Document Files 5010F, LACBS. "Research Los Angeles County Flood Control," 165, LACDPWTL; Carey McWilliams, *Southern California: An Island on the Land* (New York: Duell, Sloan, & Pearce, 1946; reprint, Salt Lake City: Peregrine Smith Books, 1994), 98.

8. The Crespi quotes appear in Rolle, *Los Angeles,* 8; the Mission San Gabriel information is from the Department of the Interior, U.S. Geological Survey, *Floods of March 1938 in Southern California,* by Harold C. Troxell et al., Water Supply Paper, U.S. Geological Survey 844 (Washington, D.C.: United States Government Printing Office, 1938), 386; Richard Henry Dana, *Two Years Before the Mast,* quoted in Anthony F. Turhollow, *A History of the Los Angeles District, U.S. Army Corps of Engineers* (Los Angeles: U.S. Army Engineer District, 1975), 20; the "clod-whacking" quote appears in Robert Glass Cleland, *The Cattle on a Thousand Hills: Southern California, 1850–1880* (San Marino, Calif.: Huntington Library, 1941), 162; the information on Browne is from "Research Los Angeles County Flood Control," 193–94, LACDPWTL.

9. On the rise of southern California's edenic image generally, see John E. Baur, *The Health Seekers of Southern California, 1870–1900* (San Marino, Calif.: Huntington Library, 1959); Kevin Starr, *Inventing the Dream: California through the Progressive Era* (New York: Oxford University Press, 1985); and McWilliams, *Southern California.* On Browne's change of heart, see "Research Los Angeles County Flood Control," 193–94, LACDPWTL. The quotes in the final sentence are from Cleland, *Cattle,* 174.

10. Thomas Balch Elliott, "History of the San Gabriel Orange Grove Association" [1875], TBE. Another example of the shifting view of climate came from a Riverside newspaper editor who wrote in 1887: "Men go not to buy land but to buy lungs. . . . The boom may burst, but the climate won't. Nature runs a bank that never breaks, and she is in business down there, hence the boom." The quote appears in Baur, *Health Seekers,* 45.

11. "No Unemployed," *LAT,* 25 February 1914, sec. II, 4; McWilliams, *Southern California,* 103.

12. "Research Los Angeles County Flood Control," 31–32, LACDPWTL.

13. Ibid., 82. See also in that volume the interview with a former California governor, H. T. Gage, who indicated that he had warned people against farming the river bottoms, but was not listened to. Subsequent floods vindicated his judgment (164).

14. There is no perfect term to describe this community, which was in great transition in the second half of the nineteenth century. I have chosen to use the term *Mexican* here, as opposed to *Mexican American,* for a variety of reasons. The processes that led to the formation of a Mexican American identity clearly began as soon as the 1848 Treaty of Guadalupe Hidalgo turned the hundred thousand Mexican inhabitants of the southwestern United States into aliens

in their own lands. Although Richard Griswold del Castillo demonstrated that people of Mexican descent in Los Angeles developed a distinct culture during the nineteenth century, the exact nature of that identity was very much contested within the community. Spanish-speaking elites, for example, as David Gutiérrez notes, often referred to themselves as Spanish instead of Mexican, in order to distinguish themselves from the working-class Mexicans (*Walls and Mirrors: Mexican Americans, Mexican Immigrants, and the Politics of Ethnicity* [Berkeley: University of California Press, 1995], 33). Moreover, George Sánchez dates the solidification of a distinctively Mexican American community to the first half of the twentieth century, especially the 1930s (*Becoming Mexican American: Ethnicity, Culture and Identity in Chicano Los Angeles, 1900–1945* [New York: Oxford University Press, 1993], 11–12). Therefore, to use the term *Mexican American* to describe the community in transition during the period between 1884 and 1910, particularly given that I am here interested in that community's memory of the pre-1848 period, seems anachronistic. For detailed discussion of the social and cultural history of people of Mexican descent in Los Angeles and the Southwest, see Richard Griswold del Castillo, *The Los Angeles Barrio, 1850–1890: A Social History* (Berkeley: University of California Press, 1979); Gutiérrez, *Walls and Mirrors;* and Sánchez, *Becoming Mexican American.*

15. "Research Los Angeles County Flood Control," 2, 45, 80–81, 121, 122, LACDPWTL.

16. Griswold del Castillo, *Los Angeles Barrio,* see especially chap. 2, "Exclusion from a Developing Economy."

17. For example, the Historical Society of Southern California was formed in 1883 as part of the movement to trumpet the virtues of the region. Its founders feared that rapid growth had brought such cultural flux to southern California that common knowledge about the region's past would quickly be lost forever in the cyclone of change that swept over the region. Consequently, the society's founders endeavored to collect and preserve knowledge of the region's colorful past to go with its florid present. They also sought to record for posterity the amazing changes of the present. To fulfill this mission, the early issues of the society's publication recounted the exploits of heroic pioneers, the romance of the Spanish mission period, colorful incidents like the deadly 1872 Owens Valley earthquake, and contemporary events like the real-estate boom of the 1880s. J. J. Warner, "Inaugural Address of the President, Col. J. J. Warner," *Publications of the Historical Society of Southern California* 1 (1884): 9–18.

18. On the relationship and evolution of these two views—disdainful and romantic—of the Spanish and Mexican past, see James J. Rawls, "The California Mission as Symbol and Myth," *California History* 71 (fall 1992): 342–61, 449–51.

19. "Research Los Angeles County Flood Control," 28, LACDPWTL; on the Germains, see 32. The phrase "old Mexican" appears frequently in the testimony of the Reagan interviews of 1915; see ibid., 26, 76, 121, 513, 529. Interviewees with Spanish surnames, however, did not use the phrase. Although both groups of people appear to have heard and believed the stories, the deroga-

tory connotation that the label "Mexican" had at that time and place suggests a certain lack of seriousness with which many of the United States–born witnesses treated the tales. Also, while the people of Mexican descent accepted the stories without question, many Americans such as D. M. Cate (76) treated them only as "possible."

20. Baur, *Health Seekers,* chap. 4; Charles Dwight Willard, Santa Barbara, to Harriet Willard, 16 October 1887, CDW; Charles Dwight Willard, Santa Barbara, to Sarah Hiestand, 30 November 1887, CDW; William H. Knight and Abbot Kinney, form letter inviting easterners to Los Angeles for the American Forestry Association convention, 24 June 1899, TPL; and William H. Knight to Theodore Parker Lukens, 26 June 1899, TPL.

21. Guinn, "Exceptional Years," 33–39.

22. *De Baker v. Southern California Railway Company,* 39 *Pac. Rep.* 615 (1895). For testimony that the court heard on previous flooding of the river and for other details about the case, see "Research Los Angeles County Flood Control," 24–25, 538–48, LACDPWTL.

23. Los Angeles County Board of Supervisors, 1889–1900, Old Document Files 4750F–4805F, passim, LACBS; Board of Engineers, "Report on San Gabriel River Control," by Frank H. Olmsted, 6 October 1913; Board of Engineers, *Reports* (1915), 2; Gumprecht, *Los Angeles River,* 146–67.

24. For examples, see "Research Los Angeles County Flood Control," 23, 34–38, 41, 74, 75, LACDPWTL; "Flood Waters Real Menace," *LAT,* 26 February 1914, sec. II, 1; "Los Angeles River Long Vexes City with Tricks: Wild Career of Flood Ravages and Shifting Channel Told in Review," *LAT,* 11 May 1924; Board of Engineers, *Reports* (1915), 2; H. Hawgood and [?], Los Angeles, to Los Angeles County Board of Supervisors, 16 August 1893, Old Document Files 4775F, LACBS.

25. Board of Engineers, "Report on San Gabriel River" (1913), 1; for other examples of floods making mockeries of flood-control projects, see "Research Los Angeles County Flood Control," 41, LACDPWTL; and H. Hawgood, Los Angeles, to Los Angeles County Board of Supervisors, Los Angeles, 28 August 1889, Old Document Files 4751F, LACBS. In 1914, Santa Fe Railroad Company officials initially hesitated to embrace flood-control efforts, because "the stream bed of today may be a mile from the stream of tomorrow"; "Quick Action on Flood Problem Is Assured," *LAT,* 28 February 1914, sec. II, 1.

26. The source of this section's heading is Board of Engineers, *Reports* (1915), 214, 217.

27. Board of Engineers, *Reports* (1915), 210; Board of Engineers, "Provisional Report of Board of Engineers Flood Control Submitted to the Board of Supervisors Los Angeles County," 1914, p. 1 of "Supplement," HL; Charles F. Queenan, *The Port of Los Angeles: From Wilderness to World Port* (Los Angeles: Los Angeles Harbor Department, 1983), 53; Rolle, *Los Angeles,* 144. Two million 1900 dollars are roughly the equivalent of forty-five million 2002 dollars. A hundred million 1910 dollars translates to two billion 2002 dollars.

28. Daniel M. Berry, San Diego, to Thomas Balch Elliott, 1 September 1873, TBE.

29. Leonard Pitt and Dale Pitt, *Los Angeles A to Z: An Encyclopedia of the City and County* (Berkeley: University of California Press, 1997), 188–89, 450–51, 550–51.

30. For descriptions of the harbor site before construction, see Turhollow, *History of the Los Angeles District,* 20–21; Cleland, *Cattle,* 158; and Rolle, *Los Angeles,* 52.

31. Emma H. Adams, *To and Fro in Southern California with Sketches of Arizona and New Mexico* (Cincinnati: W.M.B.C. Press, 1887), 220.

32. Board of Engineers, *Reports* (1915), 209. Rattlesnake Island, its shape and size altered by harbor construction, was the site of what is now known as Terminal Island.

33. Queenan, *Port of Los Angeles,* 30.

34. Adams, *To and Fro,* 250, 252–53; Queenan, *Port of Los Angeles,* 13, 27.

35. For the best overview of the harbor struggle, see William F. Deverell, "The Los Angeles Free Harbor Fight," *California History* 70 (spring 1991): 12–29, 131–35. Although a book-length scholarly history of the harbor remains to be written, several sources shed additional light on it. Charles Dwight Willard, *The Free Harbor Contest at Los Angeles: An Account of the Long Fight Waged by the People of Southern California to Secure a Harbor Located at a Point Open to Competition* (Los Angeles: Kingsley-Barnes & Neuner Co., 1899), provides an account from the perspective of one of the Free Harbor League leaders. Charles Queenan's *Port of Los Angeles* provides a celebratory general overview of the construction of the harbor. Anthony Turhollow's *History of the Los Angeles District* recounts the role of the Army Corps of Engineers.

36. Queenan, *Port of Los Angeles,* 48–53; Turhollow, *History of the Los Angeles District,* 44–48.

37. Board of Engineers, *Reports* (1915), 214, 217. The "greatest single asset of this county" comment is echoed by other engineers throughout the series of county reports in response to the 1914 floods; see also ibid., 10, 170.

38. Ibid., 214. The sum spent by the federal and local governments is roughly the equivalent of 202 million 2002 dollars. Source: http://www.cjr.org/resources/inflater.asp, accessed on 28 February 2003.

39. Ibid., 213.

40. "Research Los Angeles County Flood Control," 119, LACDPWTL.

41. Board of Engineers, *Reports* (1915), 215.

42. Ibid., 216.

43. CDPW, Division of Water Rights, *San Gabriel Investigation: Analysis and Conclusions,* by Harold Conkling, Bulletin No. 7, Reports of the Division of Water Rights (Sacramento: California State Printing Office, 1929), 9, 17; LACFCD, "Tentative Report to the Board of Supervisors of the Los Angeles County Flood Control District Outlining the Work Already Done and Future Needs of Flood Control and Conservation with Tentative Estimates, Maps, Plans and Flood Pictures," by James W. Reagan, 7 February 1924, 12–15, JBL, WRCA.

44. CDPW, *South Coastal Basin,* Bulletin No. 32 (San Francisco: California State Printing Office, 1930), 14; CDPW, California Division of Water Rights,

Biennial Report (Sacramento: California State Printing Office, 1925), 93–94; LACFCD, "Tentative Report" (1924), 14–17, JBL, WRCA.

45. An engineer investigating the river in the spring of 1914 found the flow near El Monte to be about one-twentieth of what came out of the canyon; A. L. Sonderegger, "Report on the Project of the San Gabriel Canyon and Puddingstone Storage Reservoirs," 1 December 1920, 8, manuscript, LACDPWTL.

46. CDPW, Division of Water Rights, *San Gabriel Investigation* (1929), 13–14.

47. CDPW, *South Coastal Basin,* 9.

48. "Research Los Angeles County Flood Control," 9–10, LACDPWTL.

49. Board of Engineers, "Provisional Report" (1914), 1. Forty 1894 dollars are roughly the equivalent of 833 2002 dollars. Source: http://www.cjr.org/resources/inflater.asp, accessed on 28 February 2003.

50. Board of Engineers, *Reports* (1915), 6. A thousand 1914 dollars are roughly the equivalent of eighteen thousand 2002 dollars. Source: http://www.cjr.org/resources/inflater.asp, accessed on 28 February 2003.

51. Board of Engineers, "Report on San Gabriel River" (1913), 1.

52. Ibid.

53. Starr, *Inventing the Dream,* 140–41. On the history of the fruit industry in California, see Steven Stoll, *The Fruits of Natural Advantage: Making the Industrial Countryside in California* (Berkeley: University of California Press, 1998); and the articles collected in Hal S. Barron and Richard Orsi, ed., "Citriculture and Southern California," *California History* 74 (spring 1995); California State Board of Agriculture, *Statistical Report of the California State Board of Agriculture for the Year 1914* (Sacramento: California State Printing Office, 1915), 266.

54. CDPW, *San Gabriel Investigation* (1929), 22.

55. Ronald Tobey and Charles Wetherell, "The Citrus Industry and the Revolution of Corporate Capitalism in Southern California, 1887–1944," *California History* 74 (spring 1991): 6–21, 129–31; McWilliams, *Southern California,* chap. 11, "The Citrus Belt"; and Kevin Starr, *Inventing the Dream,* chap. 5, "Works, Days, Georgic Beginnings."

56. LACFCD, "Tentative Report" (1924), 8; California State Board of Agriculture, *Statistical Report for 1914,* 266.

57. Harold E. Hedger, "A Report upon a Combined Flood Control and Water Conservation System for the San Gabriel River Los Angeles County, California," B.S. thesis, University of California, Berkeley, May 1924, 11.

58. Board of Engineers, *Reports* (1915), 2, 8, 103.

59. "Research Los Angeles County Flood Control," 62, LACDPWTL.

60. Ibid., 4, 11–12, 32, 40, 45–46, 55, 62, 82, 529.

61. Board of Engineers, *Reports* (1915), 208–13.

62. On the problems that bridges and railroad beds posed for water flow, see "Research Los Angeles County Flood Control," 51, 58, 526–27, LACDPWTL; and Gumprecht, *Los Angeles River.*

63. LACFCD, *Report of J. W. Reagan Engineer Los Angeles County Flood Control District upon the Control of Flood Waters in This District by Correction of Rivers, Diversion and Care of Washes, Building of Dikes and Dams, Pro-*

tecting Public Highways, Private Property and Los Angeles and Long Beach Harbors (Los Angeles, 1917), 27–28; Clay McShane, *Down the Asphalt Path: The Automobile and the American City* (New York: Columbia University Press, 1994), 216–20; Robert M. Fogelson, *The Fragmented Metropolis: Los Angeles, 1850–1930* (Cambridge, Mass.: Harvard University Press, 1967), 94.

64. "Research Los Angeles County Flood Control," 55, LACDPWTL.

65. Board of Engineers, *Reports* (1915), 174.

66. "Research Los Angeles County Flood Control," 25, 44–45, 70, LACDPWTL.

67. Board of Engineers, *Reports* (1915), 213–14.

68. Ibid., 174. Elsewhere in the same report (213–14), the engineer Charles Leeds noted the damage that increased cultivation caused to the harbor by adding to sedimentation.

69. The source of this section's heading is the article "Unprecedented Rain Paralyzes Traffic; 1 Dead; City Streets Are Rivers; Hundreds Marooned," *LAE,* 19 February 1914.

70. Ibid., 31; "South's Most Severe Storm Claims 5; Loss Thousands," *LAE,* 21 February 1914, sec. I, 1–2; "Arroyo Dwellers, Ranchers, Flee Raging Flood," *LAE,* 22 February 1914, sec. I, 1–2.

71. "Unprecedented Rain Paralyzes Traffic; 1 Dead; City Streets Are Rivers; Hundreds Marooned," *LAE,* 19 February 1914; "Floods and the Future," *LAT,* 26 February 1914, sec. II, 4.

72. Board of Engineers, *Reports* (1915), 213–14; "2 Drowned at Covina in Wake of Big Storm," *LAE,* 20 February 1914, sec. II, 2. Subsequent chemical analysis of this debris found it similar to that of croplands five to ten miles inland; LACFCD, *Report of J. W. Reagan* (1917), 28.

73. "Two Ships Held in River Silt, Debris Chokes Up Turning Basin," *LAE,* 22 February 1914, sec. I, 2; "Los Angeles Harbor Is Filled with Debris," *LAT,* 22 February 1914, sec. I, 11.

74. "Arroyo Dwellers, Ranchers, Flee Raging Flood," *LAE,* 22 February 1914; Board of Engineers, *Reports* (1915), 4, 216–17. Ten million 1914 dollars are roughly the equivalent of 179 million 2002 dollars; 400,000 1914 dollars are roughly the equivalent of 7 million 2002 dollars. Source: http://www.cjr.org/resources/inflater.asp, accessed on 28 February 2003.

75. "No Unemployed," *LAT,* 25 February 1914; "Storm Object Lesson: City Must Prepare for Future, Says Mayor," *LAE,* 22 February 1914, sec. I, 3; "Unprecedented Rain Paralyzes Traffic; 1 Dead; City Streets Are Rivers; Hundreds Marooned," *LAE,* 19 February 1914.

76. "Research Los Angeles County Flood Control," passim, LACDPWTL; Board of Engineers, *Reports* (1915), 4.

77. In terms of volume of water, the 1914 inundation was less than two-thirds as large as the two earlier floods for which engineers had measures. See note 6 above for exact figures and source citations.

78. [Reagan] to Board of Supervisors, 11 June 1915, Old Document Files 4970F, LACBS.

79. "Floods and the Future," *LAT,* 26 February 1914.

CHAPTER 2. A CENTRALIZED AUTHORITY AND A COMPREHENSIVE PLAN

The source of the chapter title is the Los Angeles County Flood Control Association, "Minutes of Convention," 1 July 1914, Old Document Files 4834F, LACBS.

1. LACFCA, "Minutes of Convention," 1 July 1914, Old Document Files 4834F, LACBS.

2. I borrow my analysis of institutions in this paragraph and this chapter from William Graham Sumner, *Folkways: A Study of the Sociological Importance of Usages, Manners, Customs, Mores, and Morals* (1906; reprint, Boston: Ginn & Company, 1940). Sumner maintained that an institution consists of an idea and a structure. The idea stems from beliefs deeply held by society, which, as they begin to be articulated, form precepts for arranging society. The structure is the framework and social organization for carrying out the idea. See especially pp. 53–54. I depart, however, from Sumner's faith in how widely shared the beliefs are. Sumner suggested that the ideas arose almost naturally, as human responses to existing in the world, and he also treated them as deeply held and more or less universally agreed upon. In contrast, I emphasize contingency and contestation. Here I am influenced by the historian of gender Joan Wallach Scott, who argued that "politics is the process by which plays of power and knowledge constitute identity and experience"; that is, society's deeply held beliefs that Sumner posited and southern Californians' impulse to control water that I describe here are not merely products of lived experience that arise through natural responses to the world. Rather, that experience and therefore the ideas and institutions that grow out of it are themselves products of struggles among people with different visions and different powers and capacities to pursue those interests. Joan Wallach Scott, *Gender and the Politics of History* (New York: Columbia University Press, 1988), 5. For a masterful analysis of the process by which institutions of the state arise, take shape, and function through the contestation of people with unequal power, see the essays in Gilbert M. Joseph and Daniel Nugent, eds., *Everyday Forms of State Formation: Revolution and the Negotiation of Rule in Modern Mexico* (Durham: Duke University Press, 1994).

3. "Unprecedented Rain Paralyzes Traffic; 1 Dead; City Streets Are Rivers; Hundreds Marooned," *LAE*, 19 February 1914.

4. "Storm Object Lesson: City Must Prepare for Future, Says Mayor," *LAE*, 22 February 1914. Heroism was not limited to such dramatic acts, but also included many unsung efforts. Throughout the county, families opened their homes to flood refugees. Armies of volunteers laid stones and sandbags and cleared river blockages. Thousands of workers toiled around the clock to remove debris and repair bridges, streets, and railroads; "All South Rallies as $3,000,000 Rain Goes," *LAE*, 23 February 1914, sec. I, 1–2; "Whole County Busy Remedying Ravages of Storm; Arroyo Homeless Scorn Alms; Demand Safety," *LAE*, 24 February 1914, sec. I, 2.

5. "Storm Object Lesson: City Must Prepare for Future, Says Mayor," *LAE,* 22 February 1914.

6. "Whole County Busy Remedying Ravages of Storm; Arroyo Homeless Scorn Alms; Demand Safety," *LAE,* 24 February 1914; on the public demand for official response to the floods, see the various petitions from southern Californians to the Los Angeles County Board of Supervisors requesting flood-control assistance, for example, a petition signed by four hundred people in and near the town of Bell, 6 March 1914, Old Document Files, 4820F, LACBS; "Quick Action on Flood Problem Is Assured," *LAT,* 28 February 1914; "Three Million Dollars Price of Safe Rivers," *LAT,* 27 February 1914, sec. II, 1; "Floods and the Future," *LAT,* 26 February 1914; "Something Doing Every Minute," *LAT,* 26 February 1914, sec. II, 4.

7. Scholars in the field of history of science have demonstrated that alleged scientific advances are most often as much a product of social, political, economic, and cultural conditions as they are of technical innovation. See, for examples, Eric Schatzberg, "Ideology and Technical Choice: The Decline of the Wooden Airplane in the United States, 1920–1945," *Technology and Culture* 35 (January 1994): 34–69; Carolyn Merchant, *Death of Nature: Women, Ecology, and the Scientific Revolution* (San Francisco: HarperSanFrancisco, 1980); Thomas P. Hughes, *Networks of Power: Electrification in Western Society, 1880–1930* (Baltimore: Johns Hopkins University Press, 1983); Donald C. Jackson, *Building the Ultimate Dam: John S. Eastwood and the Control of Water in the West* (Lawrence: University of Kansas Press, 1995); and James C. Williams, *Energy and the Making of Modern California* (Akron: University of Akron Press, 1996).

8. A sense of the regularity of flooding actually had existed among a few engineers since the 1911 floods had redirected the San Gabriel River into the Los Angeles Harbor. See, for example, Board of Engineers, "Report on San Gabriel River" (1913), 3. The damage occasioned by those floods, however, was not widespread enough to cause a shift in perception among most people, nor was it alarming enough to trigger the vocal demand for flood control that arose in 1914.

9. See, for example, "Research Los Angeles County Flood Control," interviews with Los Angeles County residents, compiled by James W. Reagan, 1914–1915,12–15, 16, 19, 74, 474–86, manuscript, LACDPWTL. See also D. E. Hughes, Los Angeles, to James Miller Guinn, 10 March 1922, Cave Johnson Couts Papers, HL, which contains a transcript of an interview with Guinn in October 1914.

10. Board of Engineers, "Provisional Report," 1, 4; Board of Engineers, *Reports* (1915), 172–74, 237.

11. Board of Engineers, *Reports* (1915), 237.

12. "Chop Channel to the Ocean," *LAT,* 25 February 1914.

13. Board of Engineers, *Reports* (1915), 28–30.

14. James W. Reagan, "Report upon Extent of Area and Location of Menaced Lands of Los Angeles County," 24 July 1915, 24, Old Document Files 4970F, LACBS.

15. "Storm Object Lesson: City Must Prepare for Future, Says Mayor," *LAE,* 22 February 1914.

16. Robert H. Wiebe, *The Search for Order, 1877–1920* (New York: Hill and Wang, 1967); on the widespread Progressive Era zeal for efficiency and expertise in conservation, see Samuel P. Hays, *Conservation and the Gospel of Efficiency: The Progressive Conservation Movement, 1890–1920* (Cambridge, Mass.: Harvard University Press, 1959; reprint, New York: Atheneum: 1969).

17. Hawgood to Board of Supervisors, 16 August 1893, Old Document Files 4775F, LACBS; Board of Engineers, "Report on San Gabriel River" (1913); Turhollow, *History of the Los Angeles District,* 326.

18. Los Angeles County Board of Engineers Flood Control (by Frank Olmsted), Los Angeles, to Los Angeles County Board of Supervisors, Los Angeles, 3 August 1914, Old Document Files 4838F, LACBS; Los Angeles County Board of Engineers Flood Control, Los Angeles, to Los Angeles County Board of Supervisors, Los Angeles, 19 September 1914, Old Document Files 4850F, LACBS; "Notes for River Control," n.d., Old Document Files 4857F, LACBS; Board of Engineers, *Reports* (1915), unpaginated introduction and 5–11.

19. James W. Reagan, Los Angeles, to Los Angeles County Board of Supervisors, Los Angeles, 24 July 1915, Old Document Files 4970F, LACBS; Los Angeles County Board of Engineers Flood Control (by Frank Olmsted), to Los Angeles County Board of Supervisors, Los Angeles, 13 April 1914, Old Document Files, 4826F, LACBS; "Research Los Angeles County Flood Control," LACDPWTL; Reagan, "Report upon Extent of Area," 1915, 24–27, Old Document Files 4970F, LACBS; Richard Bigger, *Flood Control in Metropolitan Los Angeles* (Berkeley: University of California Press, 1959), 117–18; H. Hawgood to James W. Reagan, 23 July 1915, Old Document Files 4968F, LACBS.

20. "Flood Control Experts Clash," *LAT,* 11 August 1915, in Clipping Book 4, p. 179, JBL.

21. Ibid.

22. "Mammoth Undertaking Fathered by County," *LAT,* 2 July 1914.

23. Board of Directors of the Chamber of Commerce of Long Beach, untitled resolution, 10 September 1914, Old Document Files 4844F, LACBS; the Board of Engineers, too, was critical of existing practice. Their report indicated that the "flood protection districts heretofore established in Los Angeles County, seven in number, have on the whole, been a disappointment and the protective works constructed have proved inadequate"; Board of Engineers, *Reports* (1915), 4.

24. "Quick Action on Flood Problem Is Assured," *LAT,* 28 February 1914; "Chop Channel to the Ocean," *LAT,* 25 February 1914.

25. On flood control, see Robert Kelley, *Battling the Inland Sea: American Political Culture, Public Policy, and the Sacramento Valley, 1850–1986* (Berkeley: University of California Press, 1989); on irrigation, see Donald J. Pisani, *From the Family Farm to Agribusiness: The Irrigation Crusade in California and the West, 1850–1931* (Berkeley: University of California Press, 1984), especially chap. 6; on immigration and tourism, see Richard J. Orsi, "Selling the Golden State: A Study of Boosterism in Nineteenth-Century California," Ph.D. diss., University of Wisconsin–Madison, 1973, especially 114–234 and 602–13.

26. Robert Kelley's *Battling the Inland Sea* details these struggles and considers them in the contexts of broader contests between America's Hamiltonian and Jeffersonian impulses. See also Robert Kelley, "Taming the Sacramento: Hamiltonianism in Action," *Pacific Historical Review* 34 (February 1965): 21–49.

27. Michael DiLeo and Eleanor Smith, *Two Californias: The Truth about the Split-State Movement* (Covelo, Calif.: Island Press, 1983), 19, 24–29, 34–40, 44; Pisani, *Family Farm,* chap. 6, "Response to Monopoly: Institutional Roots of the Irrigation District, 1868–1885"; Orsi, "Selling the Golden State," 114–234, 602–13; Roberta M. McDow, "To Divide or Not to Divide," *Pacific Historian* (fall 1996); Roberta M. McDow, "State Separation Schemes, 1907–1921," *California Historical Society Quarterly* (spring 1970); James Miller Guinn, "How California Escaped State Division," *Publications of the Historical Society of Southern California* (1905); J. P. Widney, "A Historical Sketch of the Movement for a Political Separation of the Two Californias, Northern and Southern, under Both the Spanish and American Regimes," *Publications of the Historical Society of Southern California* 1 (1888–1889).

28. Quoted in Pisani, *Family Farm,* 170.

29. Kelley, *Battling the Inland Sea,* chap. 10, 197–219.

30. In fact, in 1915, on the very day the governor of California signed the law that allowed southern Californians to tax themselves for flood control, he also vetoed a number of bills sponsored by southern legislators that contained four hundred thousand dollars in state appropriations for other southern California projects; "Pocket Veto," *LAT,* 13 June 1915, sec. I, 3.

31. "Chop Channel to the Ocean," *LAT,* 25 February 1914; "Mammoth Undertaking Fathered by County," *LAT,* 2 July 1914; Bigger, *Flood Control,* 145.

32. Kelley, *Battling the Inland Sea;* "Flood Waters Real Menace," *LAT,* 26 February 1914; "Three Million Dollars Price of Safe Rivers," *LAT,* 27 February 1914; "Mammoth Undertaking Fathered by County," *LAT,* 2 July 1914; California Debris Commission, *Flood Control—Sacramento and San Joaquin River Systems, California* (Washington, D.C., 1913), quoted in Kelley, *Battling the Inland Sea,* 230; Kelley, "Taming the Sacramento," 37–38.

33. Board of Engineers to Board of Supervisors, 19 September 1914, Old Document Files 4850F, LACBS; "Chop Channel to the Ocean," *LAT,* 25 February 1914. A frustrated member of the Flood Control Association acknowledged "the difficulties from the northern part of the state" involved with securing state financial assistance for southern California flood control. "Personally," he said, "I am a little impatient with the slow ways of government operation. It takes so long to get a bill through the legislature"; [Los Angeles County Flood Control Association], untitled meeting minutes [ca. 18 September 1914], Old Document Files 4848F, LACBS.

34. "Three Million Dollars Price of Safe Rivers," *LAT,* 27 February 1914.

35. Bigger, *Flood Control,* 13. Board of Engineers to Board of Supervisors, 19 September 1914, Old Document Files 4850F, LACBS. [LACFCA], untitled meeting minutes [ca. 18 September 1914], Old Document Files 4848F, LACBS. Although the Board of Supervisors unquestionably had the power to condemn

land, there was also some question about the extent of its general authority to execute a comprehensive flood-control program; H. C. Dillion (by A. J. Utley) to Los Angeles County Board of Supervisors, Los Angeles, 12 May 1893, Old Document Files 4770F, LACBS; Board of Engineers, *Reports* (1915), 2; Harold E. Hedger, "Harold E. Hedger Oral History Transcript: Flood Control Engineer," interview by Daniel Simms, 17, 19 May and 8 July 1965, p. 8, Oral History Program, DSCUCLA.

36. The delegates voted on their own, not as blocks representing a single city or organization. On the formation and organization of the Flood Control Association, see "Quick Action on Flood Problem Is Assured," *LAT,* 28 February 1914; Board of Supervisors, "Minutes," 18 March 1914, Old Document Files 4823F, LACBS; LACFCA, "Minutes," 1 July 1914, Old Document Files 4834F, LACBS; LACFCA, "List of Members of Los Angeles County Flood Control Association," n.d., Old Document Files 4871F, LACBS; LACFCA, "Minutes of Session," 4 March 1915, Old Document Files 4902F, LACBS; LACFCA, "Minutes of Session," 31 March 1915, Old Document Files 4912F, LACBS; LACFCA, "Minutes of Meeting," 9 June 1915, Old Documents Files 4936F, LACBS; Bigger, *Flood Control,* 13–14.

37. Supervisor Pridham, later president of the Los Angeles Chamber of Commerce, was an avowed foe of organized labor who praised Los Angeles as "the Citadel of the Open Shop"; Hamilton was a leader in the Los Angeles progressive movement. Pridham is quoted in Fogelson, *Fragmented Metropolis,* 130; see ibid., 215, for Hamilton's concerns about the struggling Progressive movement in Los Angeles.

38. For lists of delegates and member organizations, see LACFCA, "Minutes," Old Document Files 4834F, LACBS; LACFCA, "List of Members," n.d., Old Document Files 4871F, LACBS; LACFCA, "Minutes," 4 March 1915, Old Document Files 4902F, LACBS; and LACFCA, "Minutes," 9 June 1915, Old Documents Files 4936F, LACBS.

39. "Flood Waters Real Menace," *LAT,* 26 February 1914.

40. "San Gabriel Farmers Oppose Flood Plans," *LAT,* 12 March 1915, in Clipping Book 4, p. 142, JBL; LACFCA, "Minutes," 4 March 1915, Old Document Files 4902F, LACBS; LACFCA, "Minutes," 31 March 1915, Old Document Files 4912F, LACBS; Los Angeles County Board of Supervisors, untitled resolution, 12 April 1915, Old Document Files 4919F, LACBS; on the municipal financing controversy in Los Angeles generally, see Fogelson, *Fragmented Metropolis,* 94.

41. "Flood Work by Blanket Bonds," *LAT,* 31 March 1915, in Clipping Book 4, p. 155, JBL.

42. LACFCA, "Minutes," 31 March 1915, Old Document Files 4912F, LACBS; Board of Supervisors, untitled resolution, 12 April 1915, Old Document Files 4919F, LACBS. As Haas himself observed, "The members of the legislature are very much up in the air right now. They are being urged by many citizens to do various things, and it is essential that some definite plan of raising the funds shall be adopted and that the entire county get behind the plan. Unless that is done, it is probable the legislature will fail to act this session, in which event it will be two years before another bill can be presented"; "350 to For-

mulate New Measure for Flood Control," Los Angeles *Tribune,* 31 March 1915, in Clipping Book 4, p. 154, JBL.

43. Hiram Johnson to F. E. Woodley, J. J. Hamilton, C. H. Windham, and A. P. Hill, 4 June 1915, Old Document Files 4928F, LACBS; Bigger, *Flood Control,* 117; LACFCA, "Minutes," 9 June 1915, Old Document Files 4936F, LACBS; "Indorses [*sic*] Flood Control Bill," *LAT,* 10 June 1915, sec. I, 12.

44. "An Act to Create a Flood Control District," *Statutes and Amendments to the Codes of California* (1915), chap. 755, 1503–12.

45. Los Angeles County Board of Supervisors, "Proceedings of Board of Supervisors of Los Angeles County Flood Control District," 30 August 1915, pp. 13–18, Old Document Files, 5003F, LACBS.

46. On Reagan's career, see his testimony before the Federal San Gabriel River Commission; San Gabriel River Commission, "Reporter's Transcript," vol. 6, 10 March 1926, U.S. Bureau of Land Management, General Land Office, boxes 88–90, document 033824, RG 49, NAPR. See also Press Reference Library (Western Edition), *Notables of the West,* vol. 1 (New York: International News Service, 1913), 601. There is some shadiness as to why Reagan quit. The only reference to the reasons that I have found is in a Board of Supervisors debate, in which Supervisor Hamilton defended Reagan's "good reasons" for withdrawing; see Board of Supervisors, "Proceedings," 30 August 1915, p. 3, Old Document Files, 5003F, LACBS. In any case, not being a member plagued him thereafter, as critics frequently questioned his professional credentials.

47. Board of Supervisors, "Proceedings," 30 August 1915, pp. 18–20, Old Document Files, 5003F, LACBS.

48. This quote is from "Who Delays the San Gabriel Dam?" *Municipal League of Los Angeles Bulletin,* 31 July 1926, 3. For other attacks, see "A Severe Indictment," *Municipal League of Los Angeles Bulletin,* 30 April 1926, 4–5. The *Los Angeles Record* called him "a slippery politician"; see "Mr. Reagan Better Look Out!" *Los Angeles Record,* n.d., in California Scrapbook 62, p. 7, HL.

49. Epes Randolph, Tucson, to John J. Hamilton, Los Angeles, 8 August 1915, Old Document Files 4978F, LACBS.

50. Hedger, "Harold E. Hedger Oral History Transcript," 1965, p. 23, Oral History Program, DSCUCLA.

51. Bigger, *Flood Control,* 58, 115–20. This single source of agreement is something of an exception to prove the rule, because the harbor diversion plan was going to be financed by the federal government as a navigation project. Therefore, the one project that locals did not have to fund also became the one project where local conflict was minimal.

52. Reagan to Board of Supervisors, 24 July 1915, pp. 1–2, Old Document Files 4970F, LACBS.

53. A. L. Emerson, Los Angeles, to J. J. Hamilton, Los Angeles, 24 July 1915, Old Document Files 4969F, LACBS. The board's Old Document Files contain dozens more petitions listing names of several hundred people who supported Reagan's bid to become chief engineer. All petitions are similar, indicating some kind of organized effort. The signers said they had observed his work

over the last year and appreciated his personal knowledge of and attention to their particular area. They mentioned his freedom from corporate influences, his technical ability, his study of and knowledge of the petitioners' particular locality. The petitions came from Bellflower, Compton, Downey, Glendale, El Monte, Rivera, San Fernando, Van Nuys, La Crescenta, and others—all outside the city limits of Los Angeles.

54. Board of Supervisors, "Proceedings," 30 August 1915, p. 18, Old Document Files, 5003F, LACBS.

55. Los Angeles Chamber of Commerce, "Resolution in re Employment of Mr. J. W. Reagan as Flood Control Engineer," 9 September 1915, Old Document Files 5010F, LACBS; see also similar resolutions by the Los Angeles City Club and the Hollywood Board of Trade in the same folder in Old Document Files.

56. Bigger, *Flood Control,* 119. The Board of Engineers Flood Control had been created specifically to develop a plan for flood control. That work accomplished, it was disbanded, and implementation was left to the new Flood Control District.

57. Los Angeles County Board of Supervisors, untitled resolution, 20 February 1917, Old Document Files, 5344F, LACBS; Bigger, *Flood Control,* 120; On the 1916 storm, see Department of the Interior, U.S. Geological Survey, *Southern California Floods of January 1916,* by H. D. McGlashan and F. C. Ebert, Water Supply Paper, U.S. Geological Survey, 426 (Washington, D.C.: United States Government Printing Office, 1918). Four million four hundred fifty thousand 1917 dollars are roughly the equivalent of sixty-three million 2002 dollars. Source: http://www.cjr.org/resources/inflater.asp, accessed on 28 February 2003.

58. LACFCD, "Tentative Report" (1924), 2, JBL; Winston W. Crouch, *Metropolitan Los Angeles, A Study in Integration: Intergovernmental Relations,* vol. 15 (Los Angeles: Haynes Foundation, 1954), 92–93.

59. LACFCD, "Tentative Report" (1924), 3, JBL.

60. Kenneth Hewitt, *Regions of Risk: A Geographical Introduction to Disasters* (Essex, England: Addison Wesley Longman, 1997); Thomas Raymond Wellock, *Critical Masses: Opposition to Nuclear Power in California, 1958–1978* (Madison: University of Wisconsin Press, 1998); Wiebe, *Search for Order;* Hays, *Gospel of Efficiency;* Oliver Zunz, *Making America Corporate: 1870–1920* (Chicago: University of Chicago Press, 1990); Frank Fischer and Carmen Sirianni, eds., *Critical Studies in Organization and Bureaucracy* (Philadelphia: Temple University Press, 1984); Stephen Skowronek, *Building a New American State: The Expansion of the National Administrative Capacities, 1877–1920* (Cambridge: Cambridge University Press, 1982); Louis Galambos and Joseph Pratt, *The Rise of the Corporate Commonwealth: U.S. Business and Public Policy in the Twentieth Century* (New York: Basic Books, 1988); Alfred D. Chandler, Jr., *The Visible Hand: The Managerial Revolution in American Business* (Cambridge, Mass.: Belknap Press of Harvard University, 1977).

CHAPTER 3. A WEIR TO DO MAN'S BIDDING

The source of the chapter title is the newspaper article "Highest Dam in World Is Under Way Near Azusa," *LAT,* 31 October 1926, sec. V, 1.

1. Board of Engineers, *Reports* (1915), 15, 31. Sixteen million five hundred thousand 1915 dollars are roughly the equivalent of 295 million 2002 dollars. Source: see http://www.cjr.org/resources/inflater.asp, accessed on 28 February 2003.

2. "Highest Dam in World Is Under Way Near Azusa," *LAT,* 31 October 1926.

3. "Grand Canyon of Southern California to Be Site of Greatest Dam in the World," *LAE,* 4 May 1924, in California Scrapbook 62, p. 3, HL.

4. Sample ballot appearing in *LAT,* 4 May 1924, sec. II, 1. Thirty-five million 1924 dollars are roughly the equivalent of 368 million 2002 dollars. Source: http://www.cjr.org/resources/inflater.asp, accessed on 28 February 2003.

5. "Value of Flood Control," *LAT,* 26 April 1924, sec. II, 4.

6. Board of Engineers, "Report on San Gabriel River" (1913), 4–6; Bigger, *Flood Control,* 55–57; Board of Engineers, *Reports* (1915), 28–31.

7. San Gabriel River Commission, "Reporter's Transcript," vol. 6, 10 March 1926, 758, U.S. Bureau of Land Management, General Land Office, RG 49, NAPR; LACFCD, *Report of J. W. Reagan* (1917), 35.

8. LACFCD, "Tentative Report" (1924), 14, WRCA.

9. Ibid., 3; San Gabriel River Commission, "Reporter's Transcript," vol. 6, 10 March 1926, 758–61, NAPR.

10. LACFCD, "Tentative Report" (1924), 52; San Gabriel River Commission, "Reporter's Transcript," vol. 6, 10 March 1926, 760–99, NAPR.

11. Pitt and Pitt, *Los Angeles,* 18–19, 55; LACFCD, "Tentative Report" (1924), 21; LACFCD, *Report of J. W. Reagan, Chief Engineer of Los Angeles County Flood Control District upon the Control and Conservation of Flood Waters in This District by Correction of Rivers, Diversion and Care of Washes, Building of Dikes and Dams, Protecting Public Highways, Private Property and Los Angeles and Long Beach Harbors* (Los Angeles, 1926), 6; CDPW, Division of Water Rights, *Biennial Report,* 1925, 93; LACFCD, *Report of J. W. Reagan Chief Engineer of Los Angeles County Flood Control District upon the Control of Flood Waters in this District by Correction of Rivers, Diversion and Care of Washes, Building of Dikes and Dams, Protecting Public Highways, Private Property and Los Angeles and Long Beach Harbors* (Los Angeles, 1924), 4, WRCA; Crouch, *Metropolitan Los Angeles,* 91.

12. LACFCD, "Tentative Report" (1924), 31.

13. "Urge Flood Control Dam," *LAT,* 2 October 1923, sec. II, 5.

14. A. L. Sonderegger, "Report on the Project of the San Gabriel Canyon and Puddingstone Storage Reservoirs," 1 December 1920, 12, manuscript, LACDPWTL; "Report of the Flood Control Committee," 5 October 1920, 4, manuscript, LACDPWTL; CDPW, Division of Water Rights, *Biennial Report* (Sacramento: California State Printing Office, 1927), 70. Reagan was only one of many people starting to eye San Gabriel Canyon as a source of water storage. The suggestion that the waters might be tapped first surfaced in 1918, at which

point it was decided too little was known about the river's hydrology. A group called the San Gabriel Reservoir Company commissioned A. L. Sonderegger as consulting engineer to investigate the canyon. He recommended construction of a dam. In 1923, the city of Pasadena applied to the state's Division of Water Rights for permission to divert waters in the canyon. Pasadena's request helped launch a six-year state investigation to determine the amount of unappropriated discharge from the canyon. See CDPW, Division of Water Rights, *Biennial Report,* 1927, 70; Sonderegger, "Report on the Project of the San Gabriel Canyon and Puddingstone Storage Reservoirs," LACDPWTL; and CDPW, Division of Water Rights, *San Gabriel Investigation*(1929), 7.

15. "Flood Bonds Are Indorsed," *LAT,* 16 April 1924, sec. II, 8.

16. San Gabriel River Commission, "Reporter's Transcript," vol. 6, 10 March 1926, 775, NAPR. The Municipal League later disagreed with what Reagan said Davis said; see "Who Delays the San Gabriel Dam?" *Municipal League of Los Angeles Bulletin,* 31 July 1926, 1, 3.

17. Stephen R. Van Wormer, "A History of Flood Control in the Los Angeles County Drainage Area," *Southern California Quarterly* 73 (1991): 64–65.

18. "Urge Flood Control Dam," *LAT,* 2 October 1923; "To Rush Work on Flood Dam," *LAT,* 3 October 1923, sec. II, 2.

19. Rolle, *Los Angeles,* 153; CDPW, Division of Water Rights, *Biennial Report,* 1925, 98; CDPW, Division of Water Rights, *Biennial Report,* 1927, 74; "Water Saving Is Held Vital," 21 February 1924, unidentified newspaper article in California Scrapbook 62, p. 5, HL; "Reservoirs Check Drouth," unidentified newspaper article in California Scrapbook 62, p. 6, HL; "Water Conservation the Most Vital Question in Southland," *Los Angeles Times Farm and Tractor Magazine,* 4 May 1924, 1; San Gabriel River Commission, "Reporter's Transcript," vol. 6, 10 March 1926, 709–10, NAPR; "Why Flood Bonds Are Needed and Why They Won't Cost a Cent," *LAE,* 4 May 1924, California Scrapbook 62, p. 3, HL; LACFCD, "Tentative Report," 1924, 5.

20. "Urge Flood Control Dam," *LAT,* 2 October 1923.

21. "Los Angeles River Long Vexes City with Tricks," *LAT,* 11 May 1924. It should be noted that despite all the concern about the drought, the region was probably not close to experiencing a genuine shortage of water. Studies by the state's water engineer Harold Conkling revealed later in the decade that the falling water table levels were due not to overdraft but to a readjustment in the water table to compensate for the declining supply during the drought. In comparison to long-term supply, more water was still flowing into the San Gabriel Basin than people were taking out of it. What was certain, however, was that the region faced a future shortage because of growth. Conkling predicted that by 1945 demands for water would outstrip supply; CDPW, Division of Water Rights, *San Gabriel Investigation* (1929), 28.

22. LACFCD, *Report of J. W. Reagan* (1924), 8; LACFCD, "Tentative Report" (1924), 5, 27.

23. LACFCD, *Report of J. W. Reagan* (1924), 8.

24. Unidentified newspaper article [1924], in California Scrapbook 62, p. 2, HL. For more on Reagan's campaign, see "Pleads for Saving Water," 29 April 1924, unidentified newspaper article in California Scrapbook 62, p. 2, HL; "En-

gineer Urges $35,000,000 Bond for Building Dam," unidentified newspaper article in California Scrapbook 62, p. 2, HL; and "Reagan Tells of Dam Plans," unidentified newspaper article in California Scrapbook 62, p. 2, HL. Other county officials, including Supervisor Prescott Cogswell, campaigned for the measure, as did various leaders of civic organizations, such as the president of the Associated Chambers of Commerce of San Gabriel Valley, James L. Matthews; "Supervisor for Water Bond Issue," unidentified newspaper article in California Scrapbook 62, p. 2, HL; San Gabriel River Commission, "Reporter's Transcript," vol. 1, 24 February and 3 March 1926, p. 49, NAPR.

25. LACFCD, "Tentative Report" (1924), 22–24, 53.

26. "Flood Bonds Are Indorsed," *LAT,* 16 April 1924; "Favor Conservation," 1924, unidentified newspaper article in California Scrapbook 62, p. 5, HL; "Flood Control Backed by Glendora," 1924, unidentified newspaper article in California Scrapbook 62, p. 4, HL; Charles P. Sayer to Los Angeles County Board of Supervisors, Los Angeles, 14 March 1924, Old Document Files 4874F, LACBS. Sayer was secretary for the Associated Chambers of Commerce of San Gabriel Valley. His and other letters in this file (see also, for example, no. 7851F) expressed the widespread support for the measure among the San Gabriel Valley communities.

27. San Gabriel River Commission, "Reporter's Transcript," vol. 1, 24 February and 3 March 1926, p. 54, NAPR; "Water Conservation in Flood Area Explained," *LAT,* 23 March 1924, in California Scrapbook 62, p. 5, HL; "How Floods Would Be Leashed to Aid Man's Needs," *LAT,* 23 March 1924, in California Scrapbook 62, p. 5, HL; "Water Conservation the Most Vital Question in Southland," *Los Angeles Times Farm and Tractor Magazine,* 4 May 1924, 1, 13.

28. "Value of Flood Control," *LAT,* 26 April 1924.

29. "How Floods Would Be Leashed to Aid Man's Needs," *LAT,* 23 March 1924, in California Scrapbook 62, p. 5, HL.

30. "Water Conservation the Most Vital Question in Southland," *Los Angeles Times Farm and Tractor Magazine,* 4 May 1924, 13.

31. "New City Charter Wins; Police Bonds Approved," *LAT,* 7 May 1924, sec. I, 1; LACFCD, "Review of San Gabriel Case before the Federal Commission," by James W. Reagan, 29 April 1926, 39–40, LACDPWTL; San Gabriel River Commission, "Reporter's Transcript," vol. 1, 24 February and 3 March, 1926, p. 54, NAPR.

32. "New City Charter Wins; Police Bonds Approved," *LAT,* 7 May 1924.

33. "Vast Program Launched as Bonds Are Approved," *LAT,* 8 May 1924, sec. II, 1, 2.

34. LACFCD, "Tentative Report" (1924), 14.

35. Arthur Powell Davis, "Los Angeles County Flood Control and Water Conservation," 5 May 1924, Old Document Files 7932F, LACBS; George Goethals to Los Angeles County Board of Supervisors, Los Angeles, 4 October 1924, reprint in unidentified newspaper, LACDPWTL; D. C. Henny and J. D. Galloway, "Report on Forks Dam Site, San Gabriel River, Los Angeles County Flood Control District," 24 August 1924, p. 8, John Debo Galloway Papers 17, WRCA.

36. See, for examples, "Reagan and the Board of Supervisors," *Municipal League of Los Angeles Bulletin,* July 1925, 1 ("public opinion as expressed . . . by Municipal League"); "League Asks Grand Jury Probe," *Municipal League of Los Angeles Bulletin,* 30 April 1926, 1 ("We are seeking . . . to obtain for the taxpayers of this County an explanation"); "Who Delays the San Gabriel Dam?" *Municipal League of Los Angeles Bulletin,* 31 July 1926, 1 ("disinterested civic organizations," referring to itself and the California Taxpayers Association).

37. Fogelson, *Fragmented Metropolis,* chap. 11 and pp. 249–50.

38. David Woodhead, Franklin D. Howell, and Anthony Pratt to Los Angeles County Board of Supervisors, Los Angeles, 18 March 1924, Old Document Files 7882F, LACBS; Special Committee of Engineers to Executive Committee of the Municipal League, 3 April 1924, Old Document Files 7898F, LACBS; and Anthony Pratt, Los Angeles, to Los Angeles County Board of Supervisors, Los Angeles, 5 April 1924, Old Document Files 7898F, LACBS; Samuel Storrow to Los Angeles County Board of Supervisors, Los Angeles, 8 April 1924, Old Document Files, box 78, LACBS.

39. Harvey Hincks, "Comparison of Dam Sites for San Gabriel Reservoir: Report to the Municipal League of Los Angeles," October 1926, John Debo Galloway Papers 92, WRCA. For a map and graph comparing the two sites, see *Municipal League of Los Angeles Bulletin,* 28 October 1925, 8–9.

40. LACFCD, "Review of San Gabriel Case before the Federal Commission," 21, 24–25, 38, 44.

41. "An Act to Create a Flood Control District," *Statutes and Amendments to the Codes of California* (1915), chap. 755, sections 4, 15, 16.

42. Edward T. Bishop, Los Angeles, to Los Angeles County Board of Supervisors, Los Angeles, 28 August 1925, John Debo Galloway Papers 92, WRCA.

43. LACFCD, "Review of San Gabriel Case before Federal Commission," 4–5.

44. One of Pasadena's proposed alternatives recommended that the city and the Flood Control District share a single dam. The city's other proposal was the suggestion that the district could build instead at Granite Dike, a location that would not interfere with Pasadena's right-of-way. The Flood Control District, however, balked at both solutions. According to the district's calculations, the downstream site was not geologically sound; nor would it provide as much reservoir storage. And allowing Pasadena to store water in a jointly built reservoir would threaten the flood-control functions of the dam, since Pasadena did not lie in the path of San Gabriel River floods and therefore would have no incentive to release its water to make space in the reservoir in the event of a flood. San Gabriel River Commission, "Reporter's Transcript," vol. 1, 24 February and 3 March 1926, pp. 4–8, NAPR; LACFCD, "Report of the Chief Engineer of the Los Angeles County Flood Control District Comparing the Dam and the Reservoir Site at San Gabriel Forks and the Dam and Reservoir Site at Pine Canyon Containing Comparative Maps Made from Actual Surveys and Other Data," 31 July 1925, 5–7, HWRS.

45. This testimony is from a *Covina Citizen* editorial that was read into the transcript at the San Gabriel River Commission's hearings. "A Call to Action,"

Covina Citizen, in San Gabriel River Commission, "Reporter's Transcript," vol. 1, 24 February and 3 March 1926, p. 80, NAPR.

46. Ibid., pp. 62–63.

47. "Who Delays the San Gabriel Dam?" *Municipal League of Los Angeles Bulletin,* 31 July 1926; "A Severe Indictment," *Municipal League of Los Angeles Bulletin,* 30 April 1926; "Reagan's Masterpiece" *Los Angeles Record,* 11 September 1926, in California Scrapbook 62, p. 8, HL; "Reagan Unfair to San Gabriel Farmers," *Municipal League of Los Angeles Bulletin,* 20 August 1926, 1; "Graft or More Incompetence?" *Municipal League of Los Angeles Bulletin,* 23 October 1926, 7; "Reagan's Dam Bonds—Viewing the Remains," *Municipal League of Los Angeles Bulletin,* 25 November 1926, 4–5; and "Oh, Mr. Reagan, Listen Here!" *LAE,* 25 April 1927, in California Scrapbook 62, p. 9, HL.

48. "Federal Report Rebukes San Gabriel Dam . . . ," *LAT,* 12 December 1926, in California Scrapbook 62, p. 1, HL; clipping is damaged, making it impossible to read second half of the title.

49. "Reagan Gets Ready to Quit Post, Report," *LAE,* 9 November 1926, in California Scrapbook 62, p. 8, HL; "Beaty Says Time to Go," *Monrovia Post,* 22 March 1927, in California Scrapbook 62, p. 9, HL.

50. "High-Dam Need Urged at Trial," 11 January 1928, unidentified newspaper clipping in California Scrapbook 62, p. 11, HL; "Court Trumps Low Dam Card," 10 January 1928, unidentified newspaper clipping in California Scrapbook 62, p. 11, HL.

51. For a more detailed chronology of the convoluted San Gabriel Dam saga, see the dozens of newspaper articles in California Scrapbook 62, HL, especially C. H. Garrigues, "San Gabriel Dam Fiasco," *Illustrated Daily News,* series of fourteen articles beginning 1 October 1930, in California Scrapbook 62, pp. 33–35, HL; *Los Angeles County Flood Control District v. Henry W. Wright,* 213 *Cal. Rep.* 335 (1931); "Who Delays the San Gabriel Dam?" *Municipal League of Los Angeles Bulletin,* 31 July 1926, 1, 3, 5; and Bigger, *Flood Control,* 122–28.

52. "St. Francis Dam of Los Angeles Water-Supply System Fails under Full Head," *ENR,* 15 March 1928, 456; "St. Francis Dam Catastrophe—A Great Foundation Failure," *ENR,* 22 March 1928, 466–69; Pitt and Pitt, *Los Angeles,* 445; Leonard and Dale Pitt set the death toll at "more than 400." The Citizens Restoration Committee's 1928 report on the failure put the figure at 385; the Ventura County Coroner's office at 420. Charles Outland estimated between 400 and 450; Charles F. Outland, *Man-Made Disaster: The Story of St. Francis Dam* (Glendale, Calif.: Arthur Clark Company, 1977), 254. Other observers cite the hundreds of uncounted migrant farm workers who died, raising the count to over 500. See also Doyce B. Nunis, Jr., ed., *The St. Francis Dam Disaster Revisited* (Los Angeles and Ventura: Historical Society of Southern California and Ventura County Museum of History and Art, 1995).

53. On Owens Valley importation, see Abraham Hoffman, *Vision or Villainy: Origins of the Owens Valley–Los Angeles Water Controversy* (College Station, Tex.: Texas A&M Press, 1981); William L. Kahrl, *Water and Power: The Conflict over Los Angeles's Water Supply in the Owens Valley* (Berkeley: University of California Press, 1982); Marc Reisner, *Cadillac Desert: The Amer-*

ican West and Its Disappearing Water (New York: Penguin Books, 1986); John Walton, *Western Times and Water Wars: State, Culture, and Rebellion in California* (Berkeley: University of California Press, 1992); Norris Hundley, Jr., *The Great Thirst: Californians and Water, 1770s-1990s* (Berkeley: University of California Press, 1992).

54. Allan E. Sedgwick, "Geology of the Forks Dam Site San Gabriel Canyon," March 1930, 1–3, 15–16, manuscript, LACDPWTL.

55. Porter H. Albright, R. V. Meikle, and Franklin Thomas, "Report to Los Angeles County Grand Jury," 22 October 1930, 38, WRCA; see also Garrigues, "San Gabriel Dam Fiasco," *Illustrated Daily News,* in California Scrapbook 62, pp. 33–35, HL. On the history of dam-building technology, see Jackson, *Building the Ultimate Dam.*

56. San Gabriel River Commission, "Reporter's Transcript," vol. 6, 10 March 1926, 772, NAPR.

57. Albright, Meikle, and Thomas, "Report to the Los Angeles County Grand Jury," 53–54, WRCA.

58. Garrigues, "San Gabriel Dam Fiasco," *Illustrated Daily News* [2 October 1930], in California Scrapbook 62, p. 33, HL; see also "Dike Dam Flaw Exposed," 9 March 1929, unidentified newspaper article in California Scrapbook 62, p. 9, HL. Three hundred fifty-five 1925 dollars are roughly the equivalent of thirty-seven hundred 2002 dollars. Source: http://www.cjr.org/resources/inflater.asp, accessed on 28 February 2003.

59. Allan E. Sedgwick, "Report to McDonald & Kahn, Inc. on the Forks Dam Site—San Gabriel River, Los Angeles County, California," 29 June 1927, 4, CHR; Albright, Meikle, and Thomas, "Report to the Los Angeles County Grand Jury," 51, WRCA.

60. Albright, Meikle, and Thomas, "Report to the Los Angeles County Grand Jury," 60, WRCA.

61. Fred W. Heath et al., Los Angeles, to the Los Angeles County Board of Supervisors, Los Angeles, 31 December 1940, p. 2, Andrae B. Nordskog Papers 1, WRCA.

62. C. H. Garrigues, "San Gabriel Dam Fiasco," *Illustrated Daily News,* series of articles, 1–16 October 1930, in California Scrapbook 62, pp. 33–35, HL; Albright, Meikle, and Thomas, "Report to the Los Angeles County Grand Jury," 42, WRCA; Los Angeles County Grand Jury, "Resolution," 7 September 1933, Andrae B. Nordskog Papers 1, WRCA; John Anson Ford, newsletter, "Dam Suit Settlement Ordered," 14 July 1936, box 52, JAF.

63. "A Twisted Flood Control Program," *Municipal League of Los Angeles Bulletin,* 1 December 1929, 3.

64. Ibid., 4.

65. I have found no evidence to suggest Reagan did anything dishonest as chief engineer. In the extensive grand jury investigations of the corruption surrounding the San Gabriel fiasco, Reagan's name was never mentioned in connection with unethical practices. In lambasting that corruption, the muckraking *Illustrated Daily News* absolved Reagan, calling him a "rugged engineer who made mistakes but nevertheless was a devout opponent of politicians after boodle"; Garrigues, "San Gabriel Dam Fiasco," *Illustrated Daily News,* 11 Oc-

tober 1930, in California Scrapbook 62, p. 33, HL. Upon Reagan's death in 1941, the Board of Supervisors passed a resolution commending him for outstanding service to the county; Walter H. Case, ed., *Long Beach Blue Book* (Long Beach: Long Beach Press-Telegram Publishing Company, 1942), 289–90.

66. On the ordinariness of failures in technological and natural systems, see Kenneth Hewitt, "The Idea of Calamity in a Technocratic Age," in *Interpretations of Calamity from the Viewpoint of Human Ecology,* ed. Kenneth Hewitt (Boston: Allen & Unwin, 1983), 3–32; and Charles Perrow, *Normal Accidents: Living with High-Risk Technologies* (New York: Basic Books, 1984).

CHAPTER 4. A MORE EFFECTIVE SCOURING AGENT

The source of this chapter's title is the Conservation Association of Los Angeles County, "Proceedings Flood Control Conference," Los Angeles, 23 March 1934, 1, WRCA.

1. Harold E. Hedger, "Harold E. Hedger Oral History Transcript: Flood Control Engineer," interview by Daniel Simms, 17, 19 May and 8 July, 1965, pp. 20–21, Oral History Program, DSCUCLA; Charles J. Kraebel, "Study of the New Year's Storm," in "Proceedings Flood Control Conference," sponsored by Conservation Association of Los Angeles County, 23 March 1934, 3–4, WRCA; USGS, *Flood in La Cañada,* 60–61.

2. USGS, *Flood in La Cañada,* 60.

3. E. Courtlandt Eaton, "Flood and Erosion Control Problems and Their Solution," *Transactions of the American Society of Civil Engineers* 101 (1936): 1319.

4. Hedger, "Harold E. Hedger Oral History Transcript," interview, 1965, p. 31, Oral History Program, DSCUCLA.

5. Eaton, "Flood and Erosion," 1320–21.

6. Department of Agriculture, *Survey Report for the Los Angeles River Watershed* (Washington, D.C.: United States Government Printing Office, 1941), 22, PAO; Hedger, "Harold E. Hedger Oral History Transcript," interview, 1965, p. 31, Oral History Program, DSCUCLA; USGS, *Flood in La Cañada,* 82–83; "Five Guests Lose Lives as Flood Ends Gay Party," *LAT,* 2 January 1934, sec. I, 2; Elmer James Cottle, *Grandpa Remembers: Life in Montana and California, 1903–1945* (privately published by Cottle, 1990), 371, Rare Books Collection, HL.

7. Hedger, "Harold E. Hedger Oral History Transcript," interview, 1965, p. 31, Oral History Program, DSCUCLA; Eaton, "Flood and Erosion," 1319; USGS, *Flood in La Cañada,* 54; James G. Jobes, "Lessons from Major Disasters," address to the University of Southern California Institute of Government, 15 June 1939, 3, U.S. Army Corps of Engineers, PAO; E. C. Eaton, "Need for Planning in the Development of Residential Foothill Areas," in "Proceedings Flood Control Conference," sponsored by Conservation Association of Los Angeles County, 23 March 1934, 34–35, WRCA.

8. Conservation Association of Los Angeles County, "Proceedings Flood Control Conference," Los Angeles, 23 March 1934, 1, WRCA.

9. Justice B. Detwiler, *Who's Who in California: A Biographical Directory, 1928–1929* (San Francisco: Who's Who Publications, 1929), 362; Russell Holmes Fletcher, ed., *Who's Who in California: A Biographical Reference Work of Notable Living Men and Women of California,* vol. *1942–1943* (Los Angeles: Who's Who Publications Company, 1941), 272.

10. Hedger, "Harold E. Hedger Oral History Transcript," interview, 1965, p. 23, Oral History Program, DSCUCLA.

11. LACFCD, "Comprehensive Plan for Flood Control and Conservation: Present Conditions and Immediate Needs," September 1931, 7, WRCA.

12. Hedger, "Harold E. Hedger Oral History Transcript," interview, 1965, pp. 29–30, Oral History Program, DSCUCLA.

13. LACFCD, "Comprehensive Plan: Present Conditions" (1931), 5, 51–60, WRCA. According to the Army Corps district engineer, "By 1930 it became apparent that the program of construction was barely keeping pace with the increase in storm water run-off"; Theodore Wyman, Jr., "Flood Control Progress in Los Angeles County," address to the Planning Conference of the County of Los Angeles on Conservation, 9 June 1939, 1, PAO.

14. LACFCD, "Comprehensive Plan: Preliminary Outline of Ultimate Plan," 10 September 1930, 1, 6, LACDPWTL; LACFCD, "Comprehensive Plan: Present Conditions" (1931), 3, 5, 11, 42; Los Angeles County Flood Control District, "Comprehensive Plan for Flood Control and Water Conservation Assembled from Computations and Data Made prior to February 18, 1935," 1935, 10, LACDPWTL; Bigger, *Flood Control,* 59.

15. Wyman, "Flood Control Progress," 1, PAO; LACFCD, "Comprehensive Plan: Present Conditions" (1931), 1–2.

16. In 1931, the state supreme court ruled that

> the board [of Supervisors] was authorized to modify the plans and specifications with reference to this improvement not only because the construction in accordance with the original plan would have been unlawful, but also because the discovery of the actual conditions constituted such a change in conditions as would justify a modification of the plans and a relocation of the means to be employed for checking and impounding the flood waters. (*Los Angeles County Flood Control District v. Henry Wright,* 213 *Cal. Rep.* 335 [1931])

17. Hedger, "Harold E. Hedger Oral History Transcript," interview, 1965, pp. 10–11, Oral History Program, DSCUCLA; USGS, *Flood in La Cañada,* 94.

18. Reagan, "Report upon Extent of Area," 1915, Old Document Files 4970F, LACBS; Eaton, "Flood and Erosion," 1347; LACFCD, "Report on Present Conditions of District Check Dams" [ca. 1937], pp. 2–3, manuscript, LACDPWTL; LACFCD, "Comprehensive Plan: Present Conditions" (1931), 40; J. B. Lippincott, "Discussion," in Eaton, "Flood and Erosion," 1347; LACFCD, "Report to E. C. Eaton, Chief Engineer, of a Theoretical Study of the Value of Small Check Dams as Flood Control Structures," by N. B. Hodgkinson, 1 December 1930, 26, manuscript, LACDPWTL.

19. LACFCD, "Comprehensive Plan: Present Conditions" (1931), 40–41. On Eaton's reservations about check dams, see also "Conclusions of E. C. Eaton, Chief Flood Control Engineer," *Municipal League of Los Angeles Bulletin,* 1 February 1930, 6.

20. See especially the 1915 Board of Engineers report's discussion of flood control in the mountain areas, the part of the report where debris concerns would have been most likely to arise; Board of Engineers, *Reports* (1915), 5–6, 34–36. Nor did Reagan talk much about debris flows in his discussion of mountain flood control; see Reagan, "Report upon Extent of Area," 1915, Old Document Files 4970F, LACBS. In enumerating the aspects of floods to be controlled, the Los Angeles Flood Control Association Resolution of 1 July 1914, named debris in the harbor as a problem but did not specifically name foothill debris flows. To whatever extent the association did consider foothill debris flows a problem, it was not their central concern; see LACFCA, "Minutes," 1 July 1914, 1–4, Old Document Files 4834F, LACBS.

21. LACFCD, "Report on Present Conditions" [ca. 1937], 1, LACDPWTL.

22. Department of Agriculture, Field Flood Control Coordinating Committee No. 18, Appendix H-1, "The Brand Park Project," in "Volume I of Appendices: Appendices A to M to Accompany Survey Report of Los Angeles River" [ca. 1940], manuscript, box 239, RG 77, U.S. Army Corps of Engineers, NAPR; see also Appendix C.

23. LACFCD, "Comprehensive Plan: Preliminary Outline" (1930), 7, 16–17; Eaton, "Flood and Erosion," 1316–17, 1329; K. Q. Volk, "Notes Taken at A.S.C.E. Hydraulics Group Meeting," 5 March 1958, JBL; Paul Baumann, "Control of Flood Debris in San Gabriel Area," *Civil Engineering* 14 (April 1944): 144; Field Flood Control Coordinating Committee, Appendix H-1 "The Brand Park Project" [ca. 1940], 2–3, NAPR; LACFCD, "Comprehensive Plan: Present Conditions" (1931), 25–26.

24. Lippincott, "Discussion," in Eaton, "Flood and Erosion," 1347.

25. LACFCD, *Report of J. W. Reagan* (1926), 4–8, 14; "Symposium on Flood Control of Los Angeles County," *Municipal League of Los Angeles Bulletin,* 1 February 1930.

26. Frank Olmsted, "Theory and Practice in Flood Control," *Municipal League of Los Angeles Bulletin,* 1 February 1930, 6.

27. "Symposium on Flood Control of Los Angeles County," *Municipal League of Los Angeles Bulletin,* 1 February 1930. A year later, as the final touches were being added to the Comprehensive Plan, the Municipal League collected statements from the meeting and sent them to the Board of Supervisors urging the construction of check dams; "Check Dam Report Submitted on Behalf of the Municipal League and Cooperating Groups and Individuals," 12 January 1931, JAF.

28. LACFCD, "Comprehensive Plan: Present Conditions" (1931), 3; Lippincott, "Discussion," in Eaton, "Flood and Erosion," 1346; *Flood Control District v. Wright,* 213 *Cal. Rep.* 335 (1931). Thirty-three million 1931 dollars are roughly the equivalent of 393 million 2002 dollars. Twenty-two million 1931 dollars are roughly the equivalent of 262 million 2002 dollars. Source: http://www.crj.org/resources/inflater.asp, accessed on 28 February 2003.

29. In 1931, the Flood Control District requested $1.8 million (an amount that would have matched state contributions to Sacramento River flood control in northern California) to help launch the Comprehensive Plan, but budgetary

struggles between northern and southern legislators resulted in only a $600,000 appropriation for Los Angeles. In 1933, the state completely shut off Los Angeles flood-control funds and did not resume appropriations until 1945. Crouch, *Metropolitan Los Angeles,* 92–93; Bigger, *Flood Control,* 144–45.

30. LACFCD, "Comprehensive Plan: Present Conditions" (1931), 4.

31. LACFCD, "Report to E. C. Eaton" (1930), 23–26.

32. "Check Dam Report," 12 January 1931, Appendix IV, JAF. Twenty-five 1931 dollars are roughly the equivalent of 298 2002 dollars. Three hundred 1931 dollars are roughly the equivalent of 3,571 2002 dollars. Source: http://www.crj.org/resources/inflater.asp, accessed on 28 February 2003.

33. LACFCD, "Comprehensive Plan: Present Conditions" (1931), 41.

34. USGS, *Urban Sprawl and Flooding in Southern California,* by S. E. Rantz, U.S. Geological Survey Circular, 601-B (Washington, D.C.: U.S. Department of the Interior, 1970), B3; CDPW, "Report of Consulting Board on Safety of the Proposed San Gabriel Dam, Los Angeles County," November 1929, 2–5, WRCA; Los Angeles County Flood Control District, *Hydrology Manual* (Los Angeles: Los Angeles County Flood Control District, 1971), B-1 and B-2; and Ralph F. Wong and Alfonso Robles, Jr., "Flood-Control Facilities for Unique Flood Problems in Southern California" [ca. 1970], 2, PAO. For a beautifully written description of the geology and ecology that make the southern California canyons so prone to debris flows, see McPhee, *Control of Nature.*

35. Pitt and Pitt, *Los Angeles,* 86.

36. Shuirman and Slosson, *Forensic Engineering,* 199. On fire in southern California forests, see also; Richard Minnich, *The Biogeography of Fire in the San Bernardino Mountains: A Historical Study,* University of California Publications, Geography, vol. 28 (Berkeley: University of California Press, 1988); Stephen Pyne, *Fire in America: A Cultural History of Wildland and Rural Fire* (Princeton: Princeton University Press, 1982); and McPhee, *Control of Nature.*

37. Eaton, "Need for Planning" (1934), 35–36, WRCA; Harry F. Blaney, "Discussion," in Eaton, "Flood and Erosion," 1338–39. For an engaging description of the causes and character of debris flows in southern California, see McPhee, *Control of Nature;* Hedger, "Harold E. Hedger Oral History Transcript," interview, 1965, pp. 20–21, Oral History Program, DSCUCLA.

38. Lippincott, "Discussion," in Eaton, "Flood and Erosion," 1345.

39. USGS, *Flood in La Cañada,* 58–74; Eaton, "Flood and Erosion," 1319.

40. Hedger, "Harold E. Hedger Oral History Transcript," interview, 1965, pp. iv–8, 27, Oral History Program, DSCUCLA.

41. Ibid., 28.

42. Lippincott, "Discussion," in Eaton, "Flood and Erosion," 1346–47.

43. Pitt and Pitt, *Los Angeles,* 245; Grace Johnson Oberbeck, *History of La Crescenta–La Cañada Valleys: A Story of Beginnings Put into Writing* (Montrose, Calif.: Ledger, 1938), 8, 16, 79.

44. "Method Devised against Havoc of Rain Run-off," *LAT,* 4 March 1934, in Clipping Book 8, pp. 138–39, JBL.

45. Pitt and Pitt, *Los Angeles,* 245; Oberbeck, *History of La Crescenta–La Cañada Valleys,* 13, 16, 79; USGS, *Flood in La Cañada,* 55.

46. Eaton, "Flood and Erosion," 1303, 1306, 1361.

47. USGS, *Flood in La Cañada,* 53, 60; Conservation Association, "Proceedings Flood Control Conference," 1934, 10, 33, WRCA.

48. USGS, *Flood in La Cañada,* 94.

49. Deposition of foothill debris had previously taken place as high as twenty-eight hundred feet in elevation, but on New Year's Eve 1933–1934, the sludge stayed in the now deeper channels for a longer distance, down to twenty-one hundred feet—the elevation of Foothill Boulevard; USGS, *Flood in La Cañada,* 72–74.

50. Bigger, *Flood Control,* 3; USGS, *Flood in La Cañada,* 54. Five million 1933 dollars are roughly the equivalent of sixty-nine million 2002 dollars. Source: http://www.cjr.org/resources/inflater.asp, accessed on 28 February 2003.

51. Conservation Association of Los Angeles County, "Pathway of Progress," 1935, WRCA; Conservation Association, "Proceedings Flood Control Conference," 1934, WRCA; Jobes, "Lessons from Major Disasters," 1939, 4, PAO.

52. J. B. Lippincott, "Need for Coordinated Forest Protection and Flood Control Program," in "Proceedings Flood Control Conference," sponsored by Conservation Association of Los Angeles County, 23 March 1934, 43, WRCA.

53. Hedger, "Harold E. Hedger Oral History Transcript," interview, 1965, p. 36, Oral History Program, DSCUCLA; Eaton, "Need for Planning" (1934), 36, WRCA.

54. LACFCD, "Memorandum to E. C. Eaton, Chief Engineer on Status of La Crescenta–Montrose Flood Control Report," by J. H. Dockweiler, 28 May 1934, p. A, manuscript, LACDPWTL; Eaton, "Need for Planning" (1934), 35, 36, WRCA; LACFCD, *Report of J. W. Reagan* (1926), 14; LACFCD, "Comprehensive Plan: Preliminary Layout of Ultimate Development," map approved by E. Courtlandt Eaton, n.d., LACDPWTL; "Method Devised against Havoc of Rain Run-off," *LAT,* 4 March 1934, in Clipping Book 8, 138–39, JBL.

55. LACFCD, "Comprehensive Plan for Flood Control" (1935), 1–2.

56. In March, the Flood Control District applied to the Reconstruction Finance Corporation for a million-dollar loan, which it got, contingent on passing a bond measure. "Lack of Funds Prevents Curb on Flood Disasters," *LAT,* 19 October 1934, sec. I, 9.

57. Stanley R. Steenbock, "And Keep Your Honor Safe . . . : A Brief Overview and Affectionate Farewell to the Los Angeles County Flood Control District, 1915–1985" [1989], 37, LACDPWTL.

58. "Flood-Control Bonds," *LAT,* 30 October 1934, sec. II, 4; "The Rain," *LAT,* 19 October 1934, sec. II, 4.

59. "Dams' Cost Climbing Up," *LAT,* 9 October 1934, sec. II, 1. Note that the structure, despite its name, was actually the second of the two dams that formed the reconstituted San Gabriel Canyon project. The first, San Gabriel Dam No. 2, later renamed Cogswell Dam, had been completed in 1932.

60. "Halt on Dam Job Sought," *LAT,* 17 October 1934, sec. II, 1; "Halting the Dam," *LAT,* 18 October 1934, sec. II, 4; "Dam May Be Redesigned," *LAT,* 6 November 1934, sec. II, 3; "Dam Work Suspended," *LAT,* 10 November 1934, sec. II, 5.

61. "The Cost of Dams," *LAT,* 10 October 1934, sec. II, 4.

62. "Foothill Areas Periled as Storm Descends on City," *LAT,* 18 October 1934, sec. I, 1, 3; "Torrential Rains Leave Seven Dead; $150,000 Damage," *LAT,* 19 October 1934, sec. I, 1, 9; "Twenty-five Montrose Homes Victims in Flood," *LAT,* 19 October 1934, sec. I, 9.

63. "Complete Election Returns from City and County," *LAT,* 9 November 1934, sec. I, 2. More people voted on that measure than on any other on the ballot. The final tally was 323,264 in favor to 349,801 against.

64. B. R. Holloway, "The Need for Planning in the Development of Residential Foothill Areas," in "Proceedings Flood Control Conference," sponsored by Conservation Association of Los Angeles County, 23 March 1934, 38, WRCA.

65. John Anson Ford, *Thirty Explosive Years in Los Angeles County* (San Marino, Calif.: Huntington Library, 1961); Pitt and Pitt, *Los Angeles,* 154.

66. John Anson Ford, untitled press release, 14 December 1934, box 10, JAF.

67. See, for example, Lippincott, "Discussion," in Eaton, "Flood and Erosion," 1346, in which Lippincott said the San Gabriel Dam plans were "hurriedly presented, without mature engineering study."

68. Baker, "Discussion," in Eaton, "Flood and Erosion," 1359.

69. San Gabriel Valley Associated Chambers of Commerce, "Remove Eaton," November 1933, Harry Barnes Papers, WRCA.

70. "Flood-Control Bonds," *LAT,* 30 October 1934, sec. II, 4. For another published attack, see "Flood Control at Los Angeles," *ENR,* 31 January 1935: "The failure in the proper prosecution of the program," the editorial said,

> can be laid directly at the door of the county supervisors. . . . The lack of progress on the program . . . must be ascribed directly to the district organization, which places control in the hands of county supervisors, with all the attendant trappings of county politics. For years the engineering and technical approach . . . has been biased by political intrigue . . . and at no time thereafter was sound engineering analysis permitted to establish a dependable basis for solving the problem. (177)

71. Ford, untitled press release, 14 December 1934, box 10, JAF.

72. John Anson Ford, newsletter, 23 January 1935, p. 1, box 52, JAF; Los Angeles Engineering Council of Founder Societies, "Report of Special Engineering Committee on Flood Control," February 1934, 3, WRCA; Bigger, *Flood Control,* 150–51.

73. Ford, newsletter, 1 April 1935, box 52, JAF. Although Ford and Legg began the process to impose civil service requirements on the district in 1935, the process was not actually completed until 1939.

74. Steenbock, "And Keep Your Honor Safe" [1989], 38, LACDPWTL.

75. Ford, newsletter, 29 January 1935, box 52, JAF. See also, Ford's comments the following week that he expected "there will be no political interference with his [Howell's] administration"; Ford, newsletter, 20 February 1935, box 52, JAF.

76. Eaton, "Flood and Erosion," 1305; see, for other examples, the rest of Eaton's article. On page 1316, he titled a section heading "Erosion with Normal Water-shed Cover," implying unburned slopes are the norm and burned ones are the exception. On page 1329, he noted that "the best protection from debris is

the *normal* vegetative cover" (emphasis added); Eaton, "Flood and Erosion," 1316, 1329. Subsequent understanding among foresters has recast fire as a normal and ecologically necessary part of the southern California mountain environment. On the history of fire and forest ecology management in southern California, see Pyne, *Fire in America;* Ronald F. Lockmann, *Guarding the Forests of Southern California: Evolving Attitudes toward Conservation of Watershed, Woodlands, and Wilderness* (Glendale: Arthur H. Clark Company, 1981); and Minnich, *Biogeography of Fire.* For examples of the predominant thinking that advocated fire suppression and held that fires were a destructive aberration that interfered with the mountain ecosystem, see Abbot Kinney, *Forest and Water* (Los Angeles: Post Publishing Company, 1900); Theodore Parker Lukens, *Forestry in Relation to City Building* (Pasadena: Throop College of Technology, 1915); and William Wilcox Robinson, *The Forests and the People: The Story of the Los Angeles National Forest* (Los Angeles: Title Insurance and Trust Company, 1946).

77. Conservation Association, "Proceedings Flood Control Conference," 1934, iii, WRCA.

78. Ibid., 7.

79. Ibid., 11; in addition to Kraebel's comments, see ibid., 1–2.

80. Jobes, "Lessons from Major Disasters," 1939, 4, PAO.

81. Arthur G. Pickett, "Discussion," in Eaton, "Flood and Erosion," 1332–33. Another example of the call for more elaborate engineering of the landscape to reduce the fire was the Board of Supervisors' exploration of plans to replant the hillsides with more fire-resistant vegetation; Ford, newsletter, 29 October 1935, pp. 3–4, box 52, JAF.

82. Eaton, "Flood and Erosion," 1302–3; "Flood Control at Los Angeles," *ENR,* 31 January 1935, 177; Jobes, "Lessons from Major Disasters," 1939, 5–6, PAO. Three hundred million 1935 dollars are roughly the equivalent of four billion 2002 dollars. Source: http://www.cjr.org/resources/inflater.asp, accessed on 28 February 2003.

83. Arthur H. Frye, "Speech by Colonel Fry, Los Angeles District Engineer, Corps of Engineers, U.S. Army, to Construction Industries Committee, Los Angeles Chamber of Commerce, 9 September 1953," 4, PAO.

84. The analysis at the end of this paragraph is influenced by Perrow, *Normal Accidents,* which reasoned analogously for technical systems.

85. Donald M. Baker to John Anson Ford, Los Angeles, 13 December 1934, box 10, JAF.

86. Baker, "Discussion," in Eaton, "Flood and Erosion," 1356–59.

87. Eaton, "Flood and Erosion," 1362.

88. LACFCD, *Annual Report,* 1957–58 (Los Angeles, 1958), 3, WRCA.

89. Howell to Board of Supervisors, 22 April 1938, 6–10, LACDPWTL.

CHAPTER 5. THE SUN IS SHINING OVER SOUTHERN CALIFORNIA

The source of the chapter title is Turhollow, *History of the Los Angeles District,* 158.

1. Harold E. Hedger, "Harold E. Hedger Oral History Transcript," interview by Daniel Simms, 17, 19 May and 8 July 1965, pp. 62–64, Oral History Program, DSCUCLA.

2. Bigger, *Flood Control;* Turhollow, *History of the Los Angeles District;* Gumprecht, *Los Angeles River.*

3. California State Planning Board, "Flood-Plain Zoning: Possibilities and Legality with Special Reference to Los Angeles County, California," by Ralph B. Wertheimer, June 1942, WRCA.

4. Gumprecht, *Los Angeles River,* Davis, *Ecology of Fear,* and McPhee, *Control of Nature,* also discuss the many ways that development complicated and aggravated the flood-control problem.

5. "Why Build in Flood Control Channels?" *Municipal League Bulletin,* April 1938, 4; Olmsted Brothers and Bartholomew and Associates, *Parks, Playgrounds and Beaches for the Los Angeles Region* (Los Angeles: Olmsted Brothers and Bartholomew and Associates, 1930); Charles W. Eliot and Donald F. Griffin, *Waterlines: Key to Development of Metropolitan Los Angeles* (Los Angeles: Haynes Foundation, 1946); Planning Board, "Flood-Plain Zoning," 10–18; Spence D. Turner, "Need for Planning in the Development of Residential Foothill Areas," in "Proceedings Flood Control Conference," sponsored by Conservation Association of Los Angeles County, 23 March 1934, 32–33, WRCA. Various forms of flood hazard zoning were in also effect in the 1930s in cities in New Hampshire, Ohio, and Wisconsin; Planning Board, "Flood-Plain Zoning," 1.

6. Bigger, *Flood Control,* 74.

7. N. B. Hodgkinson to Paul Baumann, 27 October 1937; Hodgkinson to Baumann, 30 November 1937; Finley B. Laverty to Baumann, 28 October 1937; all housed in LACDPWTL.

8. James G. Jobes, "Lessons from Major Disasters," address to the University of Southern California Institute of Government, 15 June 1939, PAO; Turhollow, *History of the Los Angeles District,* 229.

9. LACFCD, "Preliminary Report on Foothill Flood of December 31st 1933," 4 January 1934, LACDPWTL; LACFCD, "Memorandum to E. C. Eaton, Chief Engineer on Status of La Crescenta–Montrose Flood Control Report," by J. H. Dockweiler, 28 May 1934, p. A, manuscript, LACDPWTL.

10. Greg Hise and William Deverell, ed., *Eden by Design: The 1930 Olmsted-Bartholomew Plan for the Los Angeles Region* (Berkeley: University of California Press, 2000), 32–33; F. C. Livingston to John Anson Ford, 1 January 1935, box 67, JAF; Arthur G. Arnoll, Los Angeles, to Herbert C. Legg, Los Angeles, 17 January 1935, box 67, JAF.

11. Olmsted Brothers et al., *Parks, Playgrounds, and Beaches,* 23.

12. Ibid., 11–16; Hise and Deverell, *Eden by Design,* 32–46. Geographer Terrence Young adds that the report suffered from internal flaws as well, namely, that the authors failed to make a nontechnical case for the plan that would have appealed to popular romantic notions of the impact of parks on public behavior and would have helped generate public support to overcome the chamber's opposition; Terrence Young, "Moral Order, Language and the Failure of the

1930 Recreation Plan for Los Angeles County," *Planing Perspectives* 16 (2001): 333–56.

13. Marc A. Weiss, *The Rise of the Community Builders: The American Real Estate Industry and Urban Land Planning* (New York: Columbia University Press, 1987), see especially chap. 4.

14. Los Angeles Chamber of Commerce, "Stenographer's Reports," 2 April 1936, pp. 8–10, box 21, LAACC. For additional examples of the chamber hesitating to endorse one aspect or other of federal flood control out of fear of supporting federal interference, see ibid., 2 February 1939, pp. 10–12; 25 April 1940, p. 1; 26 February 1942, pp. 2–4; and 29 June 1944, pp. 3–6, all in box 28. Nor did the U.S. Department of Agriculture think zoning was a practical possibility. It devoted only three paragraphs to the subject on the last page of a sixty-seven-page *Survey Report* in 1941 and concluded that "it would be very difficult to bring sufficient pressure to bear to initiate zoning regulations"; Department of Agriculture, *Survey Report for the Los Angeles River Watershed,* 67.

15. Los Angeles Chamber of Commerce, "Stenographer's Reports," 1925–1937, boxes 10–28, LAACC.

16. George W. Robbins and L. Deming Tilton, eds., *Los Angeles: Preface to a Master Plan* (Los Angeles: Pacific Southwest Academy, 1941), 87–89 and chap. 13.

17. Department of Agriculture, *Survey Report for the Los Angeles River Watershed,* 67.

18. Robbins and Tilton, *Los Angeles,* 8.

19. Theodore Wyman, Jr., "Flood Control Progress in Los Angeles County," address to the Planning Conference of the County of Los Angeles on Conservation, 9 June 1939, 2, PAO; R. E. Cruse, "Structures to Control Torrents," *ENR,* 8 July 1937, 69–70; Edwin C. Kelton, "Flood Control Program of the Army Corps of Engineers in Los Angeles District," presentation to the American Society of Civil Engineers, Sacramento Section, 23 April 1940, 1, PAO; LAD, "Flood Control in the Los Angeles District" [ca. 1939], manuscript, PAO. For a general survey of the Army Corps of Engineers' participation in Los Angeles flood control, see Turhollow, *History of the Los Angeles District,* chap. 5.

20. James G. Jobes, "The United States Engineer Department—Its Organization and Work in Southern California," address before the City and County Engineers Association of Los Angeles, California [ca. 1936], 14, PAO.

21. United States Army Corps of Engineers, Office of the Chief Engineer, "1936–1986: 50 Years of Flood Control," fact sheet, by Edward A. Greene, March 1986, 1, PAO; the fact sheet contains a copy of the text of the 1936 Flood Control Act. See also C. T. Newton and Harold E. Hedger, "Los Angeles County Flood Control and Water Conservation," *Journal of the Waterways and Harbors Division, Proceedings of the American Society of Civil Engineers* 85 ww2 (June 1959): 85; Jobes, "United States Engineer Department" [ca. 1936], PAO; Theodore Wyman, Jr., "Flood Control Duties of the Corps of Engineers, United States Army," 16 December 1938, 1–3, PAO. Another justification often invoked for federal involvement in Los Angeles flood control was the "national welfare" derived from industry and railroads along the southern California

rivers; "Extract from *Proceedings, Meeting of the Southern California Coastal Drainage Basin Committee,*" 18 October 1941, PAO. This rationale took on added importance when southern California became a center of the national defense industry after the outbreak of European war in 1939; Warren T. Hannum, "Flood Control in Southern California in Relation to National Defense," address to the National Resources Planning Board, Los Angeles, 1 March 1941, PAO.

22. Hedger, "Harold E. Hedger Oral History Transcript," interview, 1965, p. 38, Oral History Program, DSCUCLA.

23. Wyman, "Flood Control Progress," 1939, 3, PAO.

24. "Flood Situation in Los Angeles County," *Municipal League Bulletin,* March 1938, 4.

25. Ibid.; "L.A. Flood Control District Reports on March 2 Flood," *Municipal League Bulletin,* August 1938, 7; on zoning, see especially "Why Build in Flood Control Channels?" *Municipal League Bulletin,* April 1938.

26. C. H. Howell to John Anson Ford, 12 April 1938, LACDPWTL.

27. Los Angeles Chamber of Commerce, "Stenographer's Reports," 26 March 1936, pp. 4–9, box 21, LAACC.

28. Thomas Ford to John Anson Ford, 29 March 1936, box 71, JAF.

29. Los Angeles Chamber of Commerce, "Stenographer's Reports," 26 March 1936, pp. 4–9, box 21, LAACC.

30. Municipal League of Los Angeles to Board of Engineers for Rivers and Harbours [*sic*], Washington, D.C., 11 December 1939, box 10, JAF.

31. LACFCD, *Annual Report,* 1957–58, 3, WRCA.

32. More than a third of the 140,000 lineal feet of temporary channel protection along the Los Angeles River in the eastern San Fernando Valley was destroyed; "River Flood Analyzed," *LAT,* 12 March 1938, sec. I, 5; C. H. Howell, Los Angeles, to Los Angeles County Board of Supervisors, Los Angeles, 22 April 1938, LACDPWTL; Robert S. Lewis, El Monte, to Jerry Voorhis, Washington, D.C., 29 January 1945, U.S. Army Corps of Engineers, Civil Works Projects, box 451, RG 77, NAPR; CDPW, Division of Highways, "Joint Departmental Report on Performance of Revetment and Bank Protection Subsequent to 1937–38 Floods in Northern and Southern California," 1938, 4, GPS; John Anson Ford, "Official Measures of Major Importance 1934–1938, Supervisor John Anson Ford," p. 2, box 53, JAF; "Move Spurred for Added San Gabriel River Outlet," *LAT,* 8 March 1938, sec. I, 3.

33. For example, J. K. Andrich and his neighbors in the San Fernando Valley working-class suburb of Canoga Park, many of whom lived on fifty-five dollars a month in WPA wages, suffered twenty-one thousand dollars of damage when the Los Angeles River overflowed its temporary flood-control channel and deposited up to three feet of water in their homes; J. K. Andrich, Canoga Park, to Major Theodore Wyman, Jr., Los Angeles, 15 March 1938, box 233, RG 77, NAPR; LAD, "Damage from Storm of February 27–March 3, 1938: Interview Data Sheet—Flood Areas," 22 March 1938, box 233, RG 77, manuscript, NAPR; G. B. Bebout, Los Angeles, to Major Theodore Wyman, Los Angeles, 2 April 1938, box 233, RG 77, NAPR.

34. "River Flood Analyzed," *LAT,* 12 March 1938; Hedger, "Harold E.

Hedger Oral History Transcript," interview, 1965, pp. 41–43, Oral History Program, DSCUCLA; Lloyd Aldrich, "Special Report No. 5 to the Honorable Board of Public Works of the City of Los Angeles," 11 March 1938, box 233, RG 77, NAPR; LACFCD, "Conditions at District Dams following Flood of March 2, 1938," by Finley B. Laverty and N. B. Hodgkinson, 21 March 1938, LACDPWTL; Jobes, "Lessons from Major Disasters" (1939), 5, PAO; LACFCD, "Flood of March 2, 1938," 1938, 39–42, WRCA. The damage figures are from LAD, "Los Angeles County Drainage Area Review," December 1991, revised June 1992, 20; others have set the figures even higher; Bigger's figures were fifty-nine dead and sixty-two million dollars in damage, see Bigger, *Flood Control,* 3; and Troxell put it at eighty-seven dead and seventy-eight million dollars in damages; see USGS, *Floods of March 1938,* 2, 11–12. This discrepancy is probably due to the fact that the LAD review referred to figures for Los Angeles County only, while Bigger and Troxell considered larger areas of southern California. Forty million 1938 dollars are roughly the equivalent of 513 million 2002 dollars. Source: http://www.cjr.org/resources/inflater.asp, accessed on 28 February 2003.

35. "Mopping Up," *LAT,* 4 March 1938, sec. II, 4.

36. Turhollow, *History of the Los Angeles District,* 158; Harold H. Story, "Memoirs of Harold H. Story: Oral History Transcript," interview by Elizabeth I. Dixon, 1963, 794, Oral History Program, DSCUCLA; "Mopping Up," *LAT,* 4 March 1938; "Southland Starts Rebuilding," *LAT,* 5 March 1938, sec. I, 1, 4.

37. Turhollow, *History of the Los District,* 175; Newton and Hedger, "Los Angeles County Flood Control," 89.

38. LACFCD, *Biennial Report,* 1963–65 (Los Angeles, 1965), unpaginated cover letter, WRCA; Turhollow, *History of the Los Angeles District,* 187.

39. Floyd Suter Bixby, "Drainage Doctors: Modern Construction Efficiency in Los Angeles Flood Control Program," *Constructor,* May 1941, 27–28.

40. Turhollow, *History of the Los Angeles District,* 175, 181, 187; Bixby, "Drainage Doctors," 28; "Checking Torrential Floods," *ENR,* 1 July 1937, 22–24; LACFCD, *Biennial Report,* 1961–63, 10.

41. N. A. Matthias, "The Los Angeles Flood Control Project," *Military Engineer,* September 1941, 386; on flood-control basin operation, see LAD, "LACDA Update," September 1987, 9; William G. Cassidy, "Operation of Flood Control Basins U.S. Engineer District, Los Angeles, California," paper presented at the Convention of American Society of Civil Engineers, Los Angeles, 28 July 1943, PAO; LAD, "Flood Control in the Los Angeles District" [ca. 1943], manuscript, PAO.

42. USGS, *Urban Sprawl,* B4; Alwyn F. Luse, "The Storm Drain System of Los Angeles," 1980, 5, PAO.

43. "L.A. Dams Prevented Disaster," *Los Angeles Herald-Examiner,* 28 November 1965.

44. Some Army Corps personnel have emphasized the departures of federal efforts from the county's Comprehensive Plan, particularly with regard to the 1935 emergency construction. James Jobes, for example, observed that when the Army Corps first got involved in 1935, the county's plans "were found to be inadequate since they had been prepared chiefly for preliminary purposes only";

Jobes, "United States Engineer Department" [ca. 1936], 13, PAO. Anthony Turhollow, who wrote the official history of the Los Angeles District of the U.S. Army Corps of Engineers, also emphasized those departures; Turhollow, *History of the Los Angeles District,* 155. Overall, however, continuity between the county and federal plans turned out to be the rule rather than the exception. In contrast to Jobes and Turhollow, the political scientist Richard Bigger emphasized this continuity, maintaining that "the plans for flood control in Los Angeles have been basically the creation of the Flood Control District"; Bigger, *Flood Control,* 54, 60. Additionally, the *Engineering News-Record* noted that "federal operations are a continuation" of local efforts; "Checking Torrential Floods," *ENR,* 1 July 1937, 22.

45. A 1963 book on government structure in southern California even went so far as to count flood control as one of the metropolitan issues on which there was substantial consensus among the communities; Winston W. Crouch and Beatrice Dinerman, *Southern California Metropolis: A Study in Development of Government for a Metropolitan Area* (Berkeley: University of California Press, 1963), 217–19.

46. Los Angeles Chamber of Commerce, "Stenographer's Reports," 15 September 1938, p. 2, box 21, LAACC.

47. H. C. Legg to Gordon L. McDonough, Washington, D.C., 1936, box 10, JAF.

48. Thomas Ford, Washington, D.C., to John Anson Ford, Los Angeles, 29 March 1936, box 71, JAF.

49. Herbert C. Legg, Los Angeles, to California County Supervisors, 20 April 1938, box 10, JAF; H. C. Legg to Gordon L. McDonough, Washington, D.C., 1936, box 10, JAF.

50. LACFCD, *Annual Report,* 1941–42, 33; Ford, newsletter, 19 May 1936, p. 1, box 52, JAF; Bigger, *Flood Control,* 141; LAD, "Restudy of Whittier Narrows Project and Alternative Plans for Flood Control," by R. C. Hunter, December 1946, 23, 28, LACDPWTL.

51. Bigger, *Flood Control,* 115.

52. Thomas Ford, Washington, D.C., to John Anson Ford, Los Angeles, 29 March 1936, box 71, JAF.

53. Hannum, "Flood Control in Southern California in Relation to National Defense," 1941, 4, PAO.

54. Bixby, "Drainage Doctors," 27.

55. LACFCD, *Annual Report,* 1941–42, 47–54.

56. Myrl Ott, Long Beach, to District Engineer, Los Angeles, 7 December 1946; and D. E. Root, Long Beach, to District Engineer, Los Angeles, 5 December 1946; originals of both letters can be found in LAD, "Restudy of Whittier Narrows Project" 1946, Appendix V, part 3.

57. Eaton, "Flood and Erosion," 1357; Bigger, *Flood Control,* 10.

58. On the Boulder Canyon project and the Metropolitan Water District, see Norris Hundley, Jr., *Water and the West: The Colorado River Compact and the Politics of Water in the American West* (Berkeley: University of California Press, 1975); and Hundley, *Great Thirst.*

59. Wyman, "Flood Control Duties," 1938, 3, PAO; LAD, "Project Sum-

mary and Historical Sketch Whittier Narrows Flood-Control Basin Near El Monte, California," 28 February 1945, 6, manuscript, U.S. Army Corps of Engineers, Civil Works Projects, box 451, RG 77, NAPR.

60. LACFCD, *Annual Reports,* 1940–1959; LACFCD, *Biennial Reports,* 1959–1969; LACFCD, "Report on Flood of January 21–23 1943," by Finley B. Laverty, 2, WRCA; USGS, *Urban Sprawl,* B3; LACFCD, "Floods of January 15–18, 1952," by M. F. Burke, August 1952, WRCA; LAD, "Report on Flood Damage in Southern California from Floods of January 1954," 12 February 1954, 7, U.S. Army Corps of Engineers, General Administration Files, box 181, RG 77, NAPR.

61. The three storms in the 1950s combined to do about thirteen million dollars of damage. Adjusting 1938 damages to 1955 dollars, the earlier storm's price tag comes to about seventy-five million dollars, roughly the same amount of damage that flood officials estimated would have resulted from the 1950s storms, combined, had there been no flood-control works (twenty million dollars in 1952, fifty thousand dollars in 1954, and fifty-five million dollars in 1956). The dollar conversion source is ftp://ftp.bls.gov/pub/special.requests/cpi/cpiai.txt, accessed on 28 February 2003. For damage statistics regarding these floods, see LACFCD, "Floods of January 15–18, 1952," 20; LAD, "Report on Flood Damage in Southern California from Flood of 15–18 January 1952," 12 February 1952, 12–13, U.S. Army Corps of Engineers, General Administration Files, box 180, RG 77, NAPR; LAD, "Report on Flood Damage," 1954, 13; C. H. Chorpening to Thomas H. Kuchel, Washington, D.C., 9 February 1954, U.S. Army Corps of Engineers, General Administration Files, box 180, RG 77, NAPR; LAD, "Report on Storm and Flood of 25–27 January 1956 in Southern California," 2 March 1956, 7, General Administration Files, box 181, RG 77, NAPR.

62. L. L. Wise, "Is Los Angeles Ready for a Flood?" *ENR,* 17 May 1956, 50. A detailed hydrological report on the midcentury drought appeared in Department of the Interior, U.S. Geological Survey, *Water Resources of Southern California with Special Reference to the Drought of 1944–51,* by Harold C. Troxell, Water-Supply Paper, U.S. Geological Survey, 1366 (Washington, D.C.: United States Government Printing Office, 1957).

63. Jobes, "Lessons from Major Disasters" (1939), 4, PAO; Theodore Wyman, "Federal Government's Flood Control Program in Southern California," address to the University of Southern California Institute of Government, Public Engineering Section, Los Angeles, California, 16 June 1939, p. 2, CHR; USGS, *Floods of March 1938,* 2, 15; LACFCD, "Flood of March 2, 1938"; Appendix B, "Storm Frequency and Potential Storm Damage," in Department of Agriculture, Field Flood Control Coordinating Committee No. 18, "Volume I of Appendices: Appendices A to M to Accompany Survey Report of Los Angeles River" [ca. 1940], manuscript, box 239, RG 77, NAPR; Wise, "Is Los Angeles Ready for a Flood?" *ENR,* 17 May 1956, 40; Newton and Hedger, "Los Angeles County Flood Control," 85; Hedger, "Harold E. Hedger Oral History Transcript," 1965, pp. 41–47, Oral History Program, DSCUCLA; LAD, "Report on Storm and Flood," 1956, 3; LACFCD, "Floods of January 15–18, 1952," 20;

LAD, "Report on Flood Damage," 1954, table 7, 13; LACFCD, "Report on Debris Reduction Studies for Mountain Watersheds," by William R. Ferrell et al., November 1959, WRCA; James E. Slosson and James P. Krohn, "Southern California Landslides of 1978 and 1980," in *Storms, Floods, and Debris Flows . . . : Proceedings of a Symposium,* sponsored by NRC, 291; Adam Rome, *The Bulldozer in the Countryside: Suburban Sprawl and the Rise of American Environmentalism* (Cambridge: Cambridge University Press, 2001), chap. 5, 153–188.

64. LACFCD, "Floods of January 15–18, 1952," 20.

65. LAD, "Report on Flood Damage," 1952, 7; LAD, "Preliminary Report on Recent Forest Fires and Flood Damage, Los Angeles and San Bernardino Counties, California," 4 February 1954, 3, manuscript, U.S. Army Corps of Engineers, General Administration Files, box 180, RG 77, NAPR; LAD, "Report on Flood Damage," 1954, 14.

66. "Small Losses Prove Value of Dam System," *LAT,* 4 March 1938, sec. I, 7.

67. LAD, "Report on Flood Damage," 1952, 12; LAD, "Report on Flood Damage," 1954, 13; LAD, "Report on Storm and Flood," 1956, 7.

68. For an overview of the Whittier Narrows controversy, see Thomas Hoult, "The Whittier Narrows Dam: A Study in Community Competition and Conflict," M.A. thesis, Whittier College, 1948.

69. LAD, "Report to Board of Consultants on Whittier Narrows Dam and Appurtenant Structures," June 1949, I-1 through I-3, and II-4, WRCA.

70. LACFCD, "Comprehensive Plan for Flood Control" (1931), 29–30, WRCA.

71. Hoult, "Whittier Narrows Dam," 51.

72. For objections by school districts and El Monte property owners, see El Monte Flood Committee, El Monte, to Los Angeles County Board of Supervisors, Los Angeles, 3 May 1938, box 10, JAF; "Resolution of the City Council of the City of El Monte, Requesting the Board of Supervisors to Order a Detailed Cost Study of Channel Protection along the San Gabriel River, Before Considering Construction of a Dam at Whittier Narrows," 3 May 1938, box 10, JAF; "Resolution of the Board of Trustees of the El Monte School District, Protesting the Construction of a Proposed Dam and Reservoir at the Whittier Narrows, and Supporting the Resolution of the City of El Monte That Further Detailed Study Be Made for Channel Protection," 2 May 1938, box 10, JAF; "Resolution of the Board of Trustees of the Mt. View School District, Protesting the Construction of a Proposed Dam and Reservoir at the Whittier Narrows, and Supporting the Resolution of the City of El Monte That Further Detailed Study Be Made for Channel Protection," 2 May 1938, box 10, JAF; "Resolution of the Board of Trustees of the El Monte Union High School District, Protesting the Construction of a Proposed Dam and Reservoir at the Whittier Narrows, and Supporting the Resolution of the City of El Monte That Further Detailed Study Be Made for Channel Protection," 3 May 1938, box 10, JAF; Temple School District Board of Trustees, El Monte, to Los Angeles County Board of Supervisors, Los Angeles, 2 May 1938, box 10, JAF.

73. Rufus W. Putnam, Los Angeles, to Chief of Engineers, Washington, D.C., 6 March 1945, U.S. Army Corps of Engineers, Civil Works Projects, box 451, RG 77, NAPR; LAD, "Project Summary and Historical Sketch," 1945, 6.

74. Rufus W. Putnam, "The Proposed Civil Works Program of the Los Angeles Engineer District of the Corps of Engineers," address before the Construction Industries Committee of the Los Angeles Chamber of Commerce, 10 January 1945, Los Angeles, California, 13, PAO; LACFCD, *Annual Report,* 1941–42, 47–54.

75. Frank J. Safley, no title, report recommending alternatives to the proposed Whittier Narrows Reservoir project, 20 April 1938, box 10, JAF; LAD, "Project Summary and Historical Sketch," 1945, 10–15; LAD, "Restudy of Whittier Narrows Project," 1946, 2–3.

76. Hoult, "Whittier Narrows Dam," 179; Safley, untitled report, 20 April 1938, 1–3, box 10, JAF.

77. J. A. Bell to John Anson Ford, Los Angeles, 23 March 1939, box 10, JAF; Bell to Voorhis, 23 March 1940, box 10, JAF; Safley, untitled report, 20 April 1938, box 10, JAF.

78. Harold E. Hedger, Los Angeles, to Pat MacDonneil, Gardena, Calif., 21 December 1945, U.S. Army Corps of Engineers, Civil Works Projects, box 451, RG 77, NAPR.

79. Even El Monteans worried about this; Flood Committee to Board of Supervisors, 3 May 1938, JAF.

80. Wyman, "Flood Control Duties," 1938, 9–10, PAO; Arthur Maass, *Muddy Waters: The Army Engineers and the Nation's Rivers* (Cambridge, Mass.: Harvard University Press, 1951), 22–37.

81. Rufus W. Putnam, Los Angeles, to Chief Engineer, Los Angeles, 1 August 1944, U.S. Army Corps of Engineers, Civil Works Projects, box 451, RG 77, NAPR; LAD, "Preliminary Report of Revised General Plan for Whittier Narrows Flood Control Basin and the Improvement of San Gabriel River and Rio Hondo Channels, Downstream of Whittier Narrows Dam," 1 August 1944, manuscript, U.S. Army Corps of Engineers, Civil Works Projects, box 451, RG 77, NAPR. The district engineer gave a complete sketch of the history of changes to Whittier Narrows Dam design in LAD, "Project Summary and Historical Sketch," 1945, 6–9.

82. Lewis to Voorhis, 29 January 1945, U.S. Army Corps of Engineers, Civil Works Projects, box 451, RG 77, NAPR.

83. Hiram W. Johnson, Washington, D.C., to Chief of Engineers, Washington, D.C., 7 February 1945, U.S. Army Corps of Engineers, Civil Works Projects, box 451, RG 77, NAPR; R. Reybold, Washington, D.C., to Hiram W. Johnson, Washington, D.C., 15 February 1945, U.S. Army Corps of Engineers, Civil Works Projects, box 451, RG 77, NAPR; Putnam to Chief of Engineers, 6 March 1945, NAPR; LAD, "Project Summary and Historical Sketch," 1945; Long Beach Board of Water Commissioners, "Resolution No. 346: A Resolution Urging Congress to Appropriate Funds to Initiate the Construction of the Whittier Narrows Dam in Los Angeles County, California," 27 December 1945, U.S. Army Corps of Engineers, Civil Works Projects, box 451, RG 77, NAPR; H. A. Cate, Whittier, to District Engineer, Los Angeles, 6 November 1945, U.S. Army

Corps of Engineers, Civil Works Projects, box 451, RG 77, NAPR; Mrs. E. E. Wallace, El Monte, to Earl Warren, Sacramento, 15 September 1945, U.S. Army Corps of Engineers, Civil Works Projects, box 451, RG 77, NAPR; "Whittier Narrows Project Fund May Be Withdrawn," *Metropolitan Star-News and Pasadena Post,* 26 October 1945, clipping in U.S. Army Corps of Engineers, Civil Works Projects, box 451, RG 77, NAPR; LAD, "Restudy of Whittier Narrows Project," 1946, 1.

84. LAD, "Restudy of Whittier Narrows Project," 1946, 22–23.

85. Ibid., 23–24, 28.

86. United States Army Corps of Engineers, Los Angeles District, "Analysis of Design: Whittier Narrows Flood Control Basin," January 1950, III-2–III-4, LACDPWTL; various correspondence between Richard Nixon and Herman L. Perry and others, 1946–1948, RNL; Hoult, "Whittier Narrows Dam"; Irwin F. Gellman, *The Contender: Richard Nixon, the Congress Years, 1946–1952* (New York: Free Press, 1999).

87. "L.A. Dams Prevented Disaster," *Los Angeles Herald-Examiner,* 28 November 1965.

88. USGS, *Urban Sprawl,* B4.

CHAPTER 6. NECESSARY BUT NOT SUFFICIENT

The source of the chapter title is NRC, *Storms, Floods, and Debris Flows . . . : Overview and Summary,* 18–19.

1. LAD, "Appendix D: Report on Floods of January and February 1969 in Los Angeles County," December 1969, D18, D74–76, LACDPWTL; LACFCD, "Summary Report: Storms of 1969," by Larry D. Simpson, June 1969, 8, 20, 37, 49, LACDPWTL; USGS, *Urban Sprawl,* B6, B8–9. The figures for the destruction of the 1969 flood are for Los Angeles County alone. If the other southern California counties are included, the damage totals climb to sixty-two million dollars and ninety-two dead; ibid., B9. The Army Corps's figure is on the conservative side. The Flood Control District estimated that the January storm alone did thirty million dollars in damage to its facilities; LACFCD, "Summary Report," 1969, 33. Moreover, the U.S. Geological Survey estimated the Los Angeles County damage from the January storms to be similar to the estimate of the Army Corps, but the Geological Survey believed that the figure did not include the costs of lost business, emergency efforts, or depreciation of property value, which the corps had included in its calculations; LAD, "Appendix D," 1969, D27; USGS, *Urban Sprawl,* B9, B11. In general the 1938 and 1969 storms were comparable. In most places, 1938's slightly exceeded the severity of 1969's, though not everywhere. Generally the short-term intensities of the 1938 storm were greater than in 1969; however, the total precipitation in 1969 was greater. For comparisons of the two storms, see LACFCD, "Summary Report," 1969, 48–49; USGS, *Urban Sprawl,* B6; LAD, "Appendix D," 1969, D17–D18.

2. LAD, "Appendix D," 1969, unpaginated foreword. For similarly optimistic assessments of the storm, see LACFCD, "Summary Report," 1969, 33, 45.

3. LAD, "LACDA Update," September 1987, 1.

4. Ibid., 1, 4, 6–9.

5. Ibid., 11.

6. LACFCD, *Biennial Report,* 1969–71, 3, 8–9 WRCA.

7. Ibid., 12.

8. Sierra Madre, Minutes of City Council Meetings, 26 May 1970, p. 5, 25 September 1973, p. 6, 24 January 1967, p. 6, CHSM; "The City of Sierra Madre Points the Way," *Vall-e-Vents* 5, no. 11 (December 1967): 2, box 3, Sierra Club Papers, DSCUCLA; Sierra Madre (Calif.) City Council, "Draft of Ordinance 781," 1966, Sierra Club Papers, DSCUCLA; "Community Action in the West: Sierra Madre Creates a Wilderness Park," *Sunset,* May 1967, 34–35.

9. LAD, "Design Memorandum No. 3: Santa Anita Wash System General Design for Sierra Madre Wash Improvement," n.d., 1–3, PAO.

10. Stanley R. Steenbock, "And Keep Your Honor Safe . . . : A Brief Overview and Affectionate Farewell to the Los Angeles County Flood Control District, 1915–1985" [1989], 29, 39–40, LACDPWTL.

11. Sierra Club Oral History Project, "Southern Sierrans," 1976–1980, Bancroft Library and Sierra Club Papers, DSCUCLA; on the loyalty oath and racial exclusion controversies, see interviews with Thomas Amneus (pp. 11–19, 23–24, 30–35), Robert Marshall (pp. 4–10, 12–13, 36–37), and Robert Bear (pp. 32–35).

12. Sierra Club Oral History Project, "Southern Sierrans," vol. 2, interview with Irene M. Charnock, 1977, p. 14, Bancroft Library and Sierra Club Papers, DSCUCLA.

13. Sierra Club Oral History Project, "Southern Sierrans," 1976–1980, Bancroft Library and Sierra Club Papers, DSCUCLA; *Southern Sierran,* 1954–1978, box 1, Sierra Club Papers, DSCUCLA.

14. Quoted in Wellock, *Critical Masses,* 4.

15. On the history of community organizing, see Robert Fisher, *Let the People Decide: Neighborhood Organizing in America* (New York: Twayne Publishing, 1994); Robert Fisher and Peter Romanofsky, *Community Organization for Urban Social Change: A Historical Perspective* (Westport, Conn.: Greenwood Press, 1981); Robert B. Fairbanks and Patricia Mooney-Melvin, *Making Sense of the City: Local Government, Civic Culture, and Community Life in Urban America* (Columbus: Ohio State University Press, 2001).

16. Patricia Mooney-Melvin, "Before the Neighborhood Organization Revolution: Cincinnati's Neighborhood Improvement Associations, 1890–1940," in *Making Sense of the City,* ed. Fairbanks and Mooney-Melvin, 95; Fisher, *Let the People Decide,* 176.

17. Fisher, *Let the People Decide,* xiii; for example, the Mexican novelist and social critic Carlos Fuentes observed that citizen activism, in what he called the "social sector," was revolutionizing Mexican politics in the 1990s, heralding a flowering of democracy and an end to the corruption and exploitation that had been the hallmark of the government and private enterprise sectors of Mexican society throughout the century; see Carlos Fuentes, *A New Time for Mexico,* trans. Marina Gutman Castañeda and Carlos Fuentes (Berkeley: University of California Press, 1994).

18. Fisher, *Let the People Decide,* 232. Fischer and Sirianni noted that in the

1970s and early 1980s, Americans grew increasingly critical of the "rationalistic, conservative bias" of bureaucratic organizational models that dominated American management in the public and private sectors; Fischer and Sirianni, *Critical Studies,* xi.

19. Fisher, *Let the People Decide,* 132.

20. Samuel Hays coined this phrase; Samuel P. Hays, *Beauty, Health, Permanence: Environmental Politics in the United States, 1955–1985* (New York: Cambridge University Press, 1987).

21. On the environmental movement's preservationist impulses, see Stephen Fox, *John Muir and His Legacy: The American Conservation Movement* (Boston: Little, Brown and Company, 1981). On the history of its roots in postwar middle-class affluence, see Hays, *Beauty.* On its roots in postwar suburbanization, see Rome, *The Bulldozer in the Countryside.* For a view critical of the mainstream middle-class environmental movement and sympathetic to its left-leaning and social-justice wings, see Robert Gottlieb, *Forcing the Spring: The Transformation of the American Environmental Movement* (Washington, D.C.: Island Press, 1993). On the intellectual history of environmentalism, see Donald Fleming, "The Roots of the New Conservation Movement," *Perspectives in American History* 6 (1971): 3–179; and Roderick Nash, *Wilderness and the American Mind* (New Haven: Yale University Press, 1982). For a case study in cross-class postwar environmental activism, see Andrew Hurley, *Environmental Inequalities: Class, Race, and Industrial Pollution in Gary, Indiana, 1945–1980* (Durham: University of North Carolina Press, 1995). Aldo Leopold's *A Sand County Almanac with Other Essays on Conservation from Round River* (New York: Oxford University Press, 1966) and Rachel Carson's *Silent Spring* (Boston: Houghton Mifflin Company, 1962) were influential inspirations to the postwar environmental movement.

22. Thomas Amneus, "New Directions for the Angeles Chapter," 1977 (vol. 2); "Oil Resolution," *Southern Sierran* 25, no. 1 (January–February 1969): 11; "Comprehensive Smog Resolution Adopted," *Southern Sierran* (January–February 1970): 1; "Freeway Threatens Malibu Canyon," *Southern Sierran* (March 1970): 3; "Mining Renewed in Big Tujunga Wash," *Southern Sierran* (March 1970): 3; "Planners Threaten City with Urbicide," *Southern Sierran* (April 1970): 1, 3, 4; "Sierra Club and the Urban Scene," *Southern Sierran* (December 1971): 2–3; all are located in Sierra Club Papers, DSCUCLA.

23. Robert Olmos, "The River: The Mighty 'River of Angels' as It Looked to a Boy in the '30s," *Los Angeles Magazine,* August 1961, 18.

24. Elsworth Lee, "The Creek That Never Touches Land," *Westways,* September 1966, 28–29.

25. Osmundson, "How to Control the Flood Controllers," 31.

26. Sierra Club Papers, boxes 140, 144, DSCUCLA.

27. Wellock, *Critical Masses,* 4–5.

28. Federal legislation such as the National Environmental Protection Act (1969), the federal Clean Air and Clean Water laws (1970, 1972), and the Endangered Species Act (1973) required federal agencies to incorporate protections for the environment into their activities and allowed citizens and groups to bring suit against violations. At the state level, in 1975 the state Water Resources

Control Board adopted regulations allowing citizens to file complaints against improper use or diversion of public waters, guidelines that the Sierra Club used to force public hearings on at least one Flood Control District project. "Sierra Club Challenges County Flood Control District on Project," *LAT,* 30 November 1975, sec. II, 1.

29. LAD, "Initial Public Meetings," 1972, 14–15, manuscript, WRCA. After all, Roper acknowledged, "we want it [flood control] to satisfy your needs and your desires, because it is you who are paying the bills." The army's record on backing up its rhetorical support of civic involvement is mixed in the 1970s. On one hand, it publicized this meeting haphazardly and at the last minute. On the other hand, the agenda of the meeting was very citizen-friendly. Roper insisted that citizens speak first, and organizations and government representatives last. "Too often," he said, "I have been to public hearings where the private citizen had to wait until 11 or 12:00 and, in many cases got tired and went home and his voice was not heard. I do not buy that. My normal procedure is to start with the private citizens"; ibid., 22.

30. Foothill Areas Association, "News Release," December 1969, Sierra Club Papers, DSCUCLA; Sierra Madre Minutes of City Council Meetings, 25 October 1966, pp. 4–6, 22 November, pp. 7–9, CHSM.

31. Sierra Madre Minutes of City Council Meeting, 1 March 1967, p. 15, CHSM.

32. Ibid., 18.

33. LAD, "Santa Clara River and Tributaries, California, Interim Review Report for Flood Control: Newhall, Saugus & Vicinity, Los Angeles County," 15 December 1971, WRCA; "Sierra Club Challenges County Flood Control District on Project," *LAT,* 30 November 1975.

34. LAD, "Initial Public Meetings," 1972, 39–50, manuscript, WRCA.

35. Sierra Madre (Calif.) City Council, "Resolution No. 2230, A Resolution of the City Council of Sierra Madre Stating Its Position with Respect to the Proposed Improvement of Little Santa Anita Canyon," 10 May 1966, Sierra Club Papers, DSCUCLA; Sierra Madre Minutes of City Council Meetings, 14 December 1965, p. 4, 27 September 1966, pp. 10–11, 8 November 1966, p. 5, 22 November 1966, pp. 7–9, 1 March 1967, pp. 3, 12, 20, 9 July 1968, p. 2, 24 February 1970, p. 1, CHSM; Foothill Areas Association, news release, December 1969, box 140, Sierra Club Papers, DSCUCLA; Dennis Crowley to Jared Orsi, 19 January 1999, JPO; Sierra Madre (Calif.) City Council, "Resolution No. 76-28," 11 May 1976, Sierra Club Papers, DSCUCLA; LACFCD, *Biennial Report,* 1975–77, 13.

36. "Sierra Club Challenges County Flood Control District on Project," *LAT,* 30 November 1975, sec. II, 1; LAD, "Initial Public Meetings," 1972, 22, manuscript, WRCA; LACFCD, *Biennial Report,* 1975–77, 8, 14; LACDPW, "Los Angeles River Master Plan," 313. The district also established the Joint-Use Planning Unit to plan and implement recreational improvements in 1975; LACFCD, *Biennial Report,* 1973–75; LACFCD, *Biennial Report,* 1975–77, 14.

37. LAD, "Initial Public Meetings," 1972, 47, manuscript, WRCA.

38. Osmundson, "How to Control the Flood Controllers," 35–36.

39. LAD, "Alternative Proposals for Flood Control and Allied Purposes, South Fork of the Santa Clara River, Santa Clarita Valley, California," December 1975, 13, WRCA.

40. LACFCD, *Biennial Report,* 1977–79, 6–8.

41. NRC, *Storms, Floods, and Debris Flows . . . : Overview and Summary,* 14. In 2002 dollars, the storms did a combined damage of approximately $1.2 billion. Source: http://www.cjr.org/resources/inflater.asp, accessed on 28 February 2003.

42. Davis, "Rare and Unusual Postfire Flood Events," 244; NRC, *Storms, Floods, and Debris Flows . . . : Overview and Summary,* 11.

43. NRC, *Storms, Floods, and Debris Flows . . . : Overview and Summary,* 11. See also the summary's observation that "the storms of 1978 and 1980 were not a truly great series of storms"; ibid., 19.

44. Davis, "Rare and Unusual Postfire Flood Events," 255. Davis, the head of the Operations Section of the Flood Control District's Hydraulic Division, estimated the average return intervals for the worst foothill debris flows in 1978 to be between two hundred and one thousand years.

45. Pyke, "Return Periods of 1977–80 Precipitation," 80; Davis, "Rare and Unusual Postfire Flood Events," 244; Shuirman and Slosson, *Forensic Engineering,* 189; Davis estimated that two hundred thousand cubic yards of debris per square mile eroded from the slopes above Zachau Canyon, double the figure that flood controllers had been assuming since the 1940s; Davis, "Rare and Unusual Postfire Flood Events," 252. Shuirman and Slosson found similarly high rates of erosion above the site of disastrous flows in the San Gabriels at Hidden Springs; Shuirman and Slosson, *Forensic Engineering,* 200.

46. Davis, "Rare and Unusual Postfire Flood Events," 251.

47. Ibid., 252. Davis also recounted other similarly devastating flows at Rubio Wash in Alhambra and at the mountain community of Hidden Springs. For additional accounts of the flows, see Shuirman and Slosson, *Forensic Engineering,* chap. 8; and John McPhee, *The Control of Nature.*

48. John M. Tettemer, "Sediment Flow Hazards: Special Hydrologic Events," in *Storms, Floods, and Debris . . . : Proceedings of a Symposium,* sponsored by NRC, 221.

49. Tettemer, "Closing Comments on Debris and Sediment," 484.

50. The design conformed to the Army Corps's civil works construction criteria at the time; Joe Sciandrone et al., "Levee Failure and Distress, San Jacinto River Levee and Bautista Creek Channel, Riverside County, Santa Ana River Basin, California," in *Storms, Floods, and Debris Flows . . . : Proceedings of a Symposium,* sponsored by NRC, 361.

51. Kenneth L. Edwards, "Failure of the San Jacinto River Levees Near San Jacinto, California, from the Floods of February 1980," in *Storms, Floods, and Debris Flows . . . : Proceedings of a Symposium,* sponsored by NRC, 348.

52. Ibid., 351–56.

53. Sciandrone et al., "Levee Failure," 381–83.

54. J. David Rogers, Pleasant Hill, Calif., to Jared Orsi, Evanston, Ill., 2 September 1998, JPO. The Army Corps investigators discounted the possibility that the faulty materials had contributed significantly to the collapse, although the

team did acknowledge that the materials were indeed inadequate and would have failed in a larger flood; Sciandrone et al., "Levee Failure," 379–82.

55. Charles R. White, "Lake Elsinore Flood Disaster of March 1980," in *Storms, Floods, and Debris Flows . . . : Proceedings of a Symposium,* sponsored by NRC, 387–90.

56. Ibid., 390–91.

57. Ibid., 387–97; NRC, *Storms, Floods, and Debris Flows . . . : Overview and Summary,* 18.

58. NRC, *Storms, Floods, and Debris Flows . . . : Overview and Summary,* 18–19.

59. NRC, *Storms, Floods, and Debris Flows . . . : Proceedings of a Symposium.*

60. Edwards, "Failure of the San Jacinto River Levees," 356.

61. Tettemer, "Sediment Flow Hazards," 221–22.

62. NRC, *Storms, Floods, and Debris Flows . . . : Overview and Summary,* 22.

63. Ibid., 11; Dolores B. Taylor, "Flood Flows in Major Streams in Ventura County," in *Storms, Floods, and Debris Flows . . . : Proceedings of a Symposium,* sponsored by NRC, 155–56.

64. Sciandrone et al., "Levee Failure," 384.

65. NRC, *Storms, Floods, and Debris Flows . . . : Overview and Summary,* 14.

66. Tettemer, "Sediment Flow Hazards," 221.

67. NRC, *Storms, Floods, and Debris Flows . . . : Overview and Summary,* 18–19; Tettemer, "Sediment Flow Hazards," 227.

68. Ibid.

69. Ibid.

70. NRC, *Storms, Floods, and Debris Flows . . . : Overview and Summary,* 14.

71. "Worried L.A. River Managers: Control It," *California Coast and Ocean,* summer 1993, 24.

72. Eric Malnic and Henry Chu, "L.A. Region Faces Major Flood Risk, U.S. Warns," *LAT,* 25 October 1997, World Wide Web edition, http://www.latimes.com, accessed on 22 February 1999.

73. LAD, "Los Angeles County Drainage Area Review," December 1991, revised June 1992, 41–42.

74. Contrary to many homeowners' assumptions, standard homeowners insurance policies do not cover floods. Since 1968, the National Flood Insurance Program (NFIP), which is managed by the Federal Emergency Management Agency (FEMA), has made subsidized flood insurance available to individuals and businesses in communities that agree to follow federally established floodplain management guidelines (about nineteen thousand of the twenty thousand cities, towns, counties, and other political divisions within the United States authorized to conduct floodplain management within their jurisdiction). Since 1973, mortgage companies making federally backed loans in these communities are required by law to ensure that lendees within zones that FEMA designates as hundred-year floodplains maintain flood insurance. For more information on

federal flood insurance, see the NFIP web site, http://www.fema.gov/nfip/. For less technical summaries, see Sara Graham, "Flooding and Your Mortgage," *LAT,* 10 February 1997; and Marcie Geffner, "Immerse Yourself in This Primer on Flood Insurance," *LAT* home edition, 10 May 1998, sec. K, 1.

75. Douglas P. Shuit, "1,200 Protest Costly Change in Flood Maps," *LAT* home edition, 13 March 1998, sec. B, 1; Jack Leonard, "A Look Ahead," *LAT* home edition, 29 June 1998, sec. B, 1.

76. Jack Leonard, "Flood Insurance Rate Hike Is Canceled," *LAT* home edition, 3 July 1998, sec. B, 5.

77. Shuit, "1,200 Protest Costly Change in Flood Maps," *LAT* home edition, 13 March 1998; U.S. Army Corps of Engineers, "Completed LACDA Project Increases Flood Protection," http://www.spl.usace.army.mil/lacda/lacda.html, accessed on 3 May 2002; "Knabe Announces Elimination of Flood Insurance for Lower Los Angeles River Area Homeowners," press release from office of Supervisor Don Knabe, 11 January 2002, http://ladpw.org/wmd/LACDA?press_release.pdf, accessed on 3 May 2002. The quote is from Waldie, "Myth of the L.A. River," 115.

78. Jeffrey L. Rabin, "Supervisors OK Plan to Raise Walls of L.A. River," *LAT,* 4 September 1996, sec. B, 1, 6.

79. The source of the subheading is Lewis MacAdams, "Restoring the Los Angeles River: A Forty-Year Art Project," *Whole Earth Review,* spring 1995, 66.

80. City of Los Angeles, Los Angeles River Task Force, "Report of the City of Los Angeles, Los Angeles River Task Force," January 1992, 1, LAPL.

81. Ibid., 7 and Appendices A and B.

82. Note, Hansen Dam's name is misspelled in the original quotation, which is found ibid., Appendix C.

83. For a list of the ideas generated, see ibid., Appendix C.

84. Ibid., 3–4.

85. MacAdams, "Restoring the Los Angeles River," 64.

86. Ibid., 66.

87. "Friends of the L.A. River: Improve It," *California Coast and Ocean,* summer 1993, 26.

88. Prentiss Williams, "Los Angeles River: Overflowing with Controversy," *California Coast and Ocean,* summer 1993, 17; "Peter Goodwin: Consider These Ideas," *California Coast and Ocean,* summer 1993, 28–29.

89. MacAdams, "Restoring the Los Angeles River," 66 (emphasis in the original).

90. "The Los Angeles River Is: A Big Joke, a Killer in Concrete, a 50-Mile-Long Ditch, All of the Above," *Aqueduct,* spring 1978, 4–9; Dick Roraback, "Up a Lazy River, Seeking the Source: Your Explorer Follows in Footsteps of Gaspar de Portolá," *LAT,* 20 October 1985, sec. I, 1–2; and Dick Roraback, "The L.A. River Practices Own Trickle-Down Theory," *LAT,* 27 October 1985, sec. VI, 1, 6–7; Judith Coburn, "Whose River Is It, Anyway?" *Los Angeles Times Magazine,* 20 November 1994, 20, 50.

91. "Friends of the L.A. River," 27.

92. MacAdams, "Restoring the Los Angeles River," 67.

93. Coburn, "Whose River?" 50.

94. LACDPW, "Los Angeles River Master Plan," 13.

95. Ibid., 13, 17.

96. Ibid., 18.

97. Gumprecht, *Los Angeles River,* 295.

98. Christine Perla, FoLAR Technical Advisory Board Chair, to Ron Ganzfried, U.S. Army Corps of Engineers, 21 October 1991; the letter and the army's response appear in LAD, "Los Angeles County Drainage Area Review: Final Feasibility Study," 1992, EIS Appendix J, 7, 12.

99. Coburn, "Whose River?" 54.

100. The description of the meeting, including the quotes, are from Coburn, "Whose River?" 51.

101. Gottlieb, *Forcing the Spring;* Benjamin Kline, *First along the River: A Brief History of the U.S. Environmental Movement* (San Francisco: Acada Books, 2000), 118–19; Laura Pulido, "Multiracial Organizing among Environmental Justice Activists in Los Angeles," in *Rethinking Los Angeles,* ed. Michael J. Dear, H. Eric Schockman, and Greg Hise (Thousand Oaks, Calif.: Sage Publications, 1996), 171–88; the Newman quotes are from Gottlieb, *Forcing the Spring,* 166, 169.

102. Chi Mui, telephone interview by author, 15 May 2002.

103. Gottlieb, *Forcing the Spring,* 260.

104. Mui interview; Jill Stewart, "Forget It, Jake. It's Chinatown," New Times Los Angeles Online, 30 March–5 April 2000, http://www.newtimesla.com/issues/current/stewart.html, accessed on 31 March 2000; Bill Donahue, "The Same River Twice," *Mother Jones,* December 2000.

105. Mui interview; Lewis MacAdams, telephone interview by author, 15 May 2002.

106. Mui interview; MacAdams interview; Fisher, *Let the People Decide;* Jennifer Price, "Paradise Reclaimed: A Field Guide to the L.A. River," *L.A. Weekly,* 10–16 August 2001, 23–39.

107. "New Cornfield Proposal Includes Canal and School," *LAT,* 20 April 2001; MacAdams interview.

108. Coburn, "Whose River?" 51.

109. Price, "Paradise Reclaimed," 29.

110. Ibid.; MacAdams interview; Friends of the San Gabriel River web site, "Job Announcement," http://www.sangabrielriver.org, accessed on 15 May 2002; Taylor Yard web site, http://tayloryard.org/, accessed on 20 May 2002. During the early years of the 1990s, the San Gabriel River was given considerably less attention than its sister stream. While the Los Angeles River coursed past many of the markers of regional identity—the "Valley," several movie studios, Griffith Park, Dodger Stadium, Union Station, Chinatown, East L.A., downtown, the region's industrial heart, and the harbor—the San Gabriel watershed embraced more than fifty nondescript and mostly working-class suburbs. The 1980 storm did not top its banks. FEMA did not try to classify its environs as a hazard zone. It was never proposed for freeway conversion. It lacked the artistic bridges that spanned the Los Angeles River. And it had never been the setting for a blockbuster movie. For all of these reasons, the San Gabriel

River did not make as hip a cause as the Los Angeles did. By the turn of the twenty-first century, however, the San Gabriel began to garner more attention. The "Re-Envisioning the San Gabriel River Conference" was held in December 1999, and Friends of the San Gabriel River formed in 2000. Most important, the state legislature created the San Gabriel and Lower Los Angeles Rivers and Mountains Conservancy to prepare the Parkways and Open Space Plan, the first phase of which got underway in the fall of 2001.

111. James Ricci, "Snapshots from the Center of the Universe; A Park with No Name (Yet) but Plenty of History," *Los Angeles Times Magazine,* 15 July 2001, 5.

EPILOGUE. THE HISTORICAL STRUCTURE OF DISORDER

1. Robert Lee Hotz and Frank Clifford, "A Glitch in the System," *LAT,* 14 August 1996, sec. A, 1; Adam Pertman, "Power Outage Strikes 7 Western States," *Boston Globe,* 11 August 1996, sec. B, 2; Bruce Haring, "Power Outages Give PC Owners an Unpleasant Jolt," *USA Today,* 19 August 1996, sec. D, 3; Petula Dvorak, "Manholes Erupt Again in Georgetown," *Washington Post,* 14 September 2000, sec. B, 3; Robert Tait, "Georgetown's Cast Iron Excuse for a Crisis," *Scotsman,* 16 June 2001, 14; Stephen Braun, "Hungry for Power in Georgetown," *LAT,* 16 June 2001, sec. A, 10; Clarence Williams and David A. Fahrenthold, "Manhole Fire Shuts Down Georgetown," *Washington Post,* 14 July 2001, sec. B, 1; Dan Fesperman, "With a Rumble, Chaos," *Baltimore Sun,* 21 July 2001, sec. A, 1; Scott Calvert, "Plans to Repair Water Main Hit Another Snag," *Baltimore Sun,* 21 July 2001, sec. A, 6; Heather Dewar, Marcia Myers, and Kimberly A. C. Wilson, "Accident Plan Leaves City Unprepared," *Baltimore Sun,* 26 July 2001, sec. A, 1; "Preparing for Failure by Failing to Prepare," *Baltimore Sun,* 27 July 2001, sec. A, 22; Paul Adams, "Tunnel Blame Drama Begins," *Baltimore Sun,* 27 July 2001, sec. C, 1; "Water Main Studied in Train Derailment," *New York Times,* 24 July 2001, www.nytimes.com; "Chemical in Manhole Explosions Linked to CSX Train Derailment," Associated Press State and Local Wire, 7 September 2001, accessed online via Lexis-Nexis; Stephen Kiehl, "Fact-Finding Phase of Probe into Train Tunnel Fire Nears End," *Baltimore Sun,* 10 December 2002, sec. B, 10; Harvard School of Public Health, "Mosquito Borne Viruses," http://www.hsph.harvard.edu/mosquito/facts.html, accessed on 17 February 2003; "Stalking the Common Mosquito—and a Deadly Virus," National Public Radio, http://www.npr.org/programs/morning/features/2001/jul/010713.mosquito.html, accessed on 17 February 2003.

2. Char Miller, "Flood of Memories," *Texas Observer,* 27 September 1996, 16–17; Lewis F. Fisher, *Saving San Antonio: The Precarious Preservation of a Heritage* (Lubbock, Tex.: Texas Tech University Press, 1996); Lewis F. Fisher, *Crown Jewel of Texas: The Story of San Antonio's River* (San Antonio: Maverick Publishing Company, 1997).

3. Joe Shoemaker, *Returning the Platte to the People: A Story of A Unique Committee, the Platte River Development Committee* (Westminster, Colo.: Greenway Foundation, 1981); Denver Urban Renewal Authority, "Economic

Feasibility Study for the Denver Platte River Valley," by Reuben A. Zubrow, 7 June 1966, University of Colorado, Boulder; Jeff Miller, "Denver's New Front Yard—Good Times Down by the Riverside," *Boston Globe,* 3 June 2001, sec. F, 16; Wellington E. Webb, "Restoring Denver's South Platte River," *Urban Land* 57 (April 1998): 16; Phil Goodstein, *Denver in Our Time,* vol. 1: *Big Money in the Big City* (Denver: New Social Publications, 1999).

4. Liz Sly, "Three Gorges Dam: Great Leap of Faith," *Chicago Tribune,* 20 December 1996, sec. 1, 1, 11; Indira A. R. Lakshmanan, "China's Pride and Pain: Three Gorges Dam Is a Monument to Some, an Endless Fountain of Controversy to Others," *Boston Globe,* 6 November 2000, sec. A, 1; Peter N. Spotts, "It Could Get Hotter in Japan Thanks to Three Gorges Dam," *Christian Science Monitor,* 12 April 2001, Features section, 14.

5. George Bugliarello, "Biology, Society and Machines," *American Scientist,* May-June 1998, 232.

6. Robert Tait, "Georgetown's Cast Iron Excuse for a Crisis," *Scotsman,* 16 June 2001, 14; Stephen Braun, "Hungry for Power in Georgetown," *Los Angeles Times,* 16 June 2001, sec. A, 10.

7. "Fiddling While Parks Burn," *Denver Post,* 14 June 2001, sec. B, 6; "Report Rejects Disciplining of Parks Officials in Wildfire," *New York Times,* 13 June 2001, sec. A, 28.

8. "Transcription of Press Conference Regarding the Cerro Grande Fire Held May 18, 2000 in Sante [*sic*] Fe, New Mexico, with Secretary Bruce Babbitt," 26 May 2000, http://www.nps.gov/cerrogrande/transcript.htm; Tony Davis, "The West's Hottest Question: How to Burn What's Bound to Burn," *High Country News,* 5 June 2000; William deBuys, "Los Alamos Fire Offers a Lesson in Humility," *High Country News,* 3 July 2000.

9. U.S. Department of the Interior, National Parks Service, Board of Inquiry, "Cerro Grande Prescribed Fire: Board of Inquiry Final Report," 26 February 2001, 14–16, 22, 26, http://www.nps.gov/fire/, accessed on 13 June 2001; U.S. Department of Interior, U.S. National Interagency Fire Center, Fire Investigation Team, "Cerro Grande Prescribed Fire, May 4–8, 2000, Investigation Report," 18 May 2000, http://www.nps.gov/cerrogrande/, accessed on 13 June 2001.

10. Board of Inquiry, "Cerro Grande Prescribed Fire," i, ii.

11. "Mother Nature's 'Great Mischief,'" *Chicago Tribune,* 27 August 1998, sec. I, 22; Davis, *Ecology of Fear,* 65–72; McPhee, *The Control of Nature.*

12. Stephen Jay Gould, *Wonderful Life: The Burgess Shale and the Nature of History* (New York: W. W. Norton, 1989), 277–91, 301, 305–8. There is some controversy today about the nature of Darwinian evolution. For a contrasting approach to Gould's, see Daniel Dennet, *Darwin's Dangerous Idea: Evolution and the Meanings of Life* (New York: Simon and Schuster, 1995). For Gould's rebuttal, see Stephen Jay Gould, "Darwinian Fundamentalism," *New York Review of Books,* 12 June 1997, 34–37; and Stephen Jay Gould, "Evolution: The Pleasures of Pluralism," *New York Review of Books,* 26 June 1997, 47–52.

13. Gould, *Wonderful Life,* 45–52.

14. Donald Worster, *Nature's Economy: A History of Ecological Ideas,* 2d ed. (New York: Cambridge University Press, 1990); Frederic E. Clements,

Plant Succession: An Analysis of the Development of Vegetation (Washington, D.C.: Carnegie Institution of Washington, 1916); Barbour, "Ecological Fragmentation in the Fifties"; and Ronald C. Tobey, *Saving the Prairies: The Life Cycle of the Founding School of American Plant Ecology, 1895–1955* (Berkeley: University of California Press, 1981); William Cronon, *Changes in the Land: Indians, Colonists, and the Ecology of New England* (New York: Hill and Wang, 1983), 27; Arthur F. McEvoy, "Toward an Interactive Theory of Nature and Culture: Ecology, Production, and Cognition in the California Fishing Industry," *Environmental Review* 11 (winter 1987): 289–305.

15. Gould, *Wonderful Life,* 283.

16. Knudson and Vogel, "Superfloods Threaten West."

17. Ilya Prigogine and Isabelle Stengers, *Order out of Chaos: Man's New Dialogue with Nature* (Boulder, Colo.: New Science Library, 1984); James Gleick, *Chaos: Making a New Science* (New York: Penguin Books, 1987).

18. Perrow, *Normal Accidents;* on complexity and coupling generally, see 3–12; on nuclear power, see 15–31; on missed job interview, see 5–6; on airline electrical accidents, see 134–35.

19. J. David Rogers, Pleasant Hill, California, to the author, Evanston, Illinois, 2 September 1998, typed letter in author's possession.

20. My analysis in this paragraph is influenced by the sociologist Anthony Giddens's theory of "structuration"; Anthony Giddens, *The Constitution of Society: Outline of the Theory of Structuration* (Berkeley: University of California Press, 1984). Society, Giddens argued, is constituted by the interaction of everyday activities and the social structures within which they occur. Daily activity of individuals, he maintained "draws upon and reproduces structural features of wider social systems" (24). Those structures both enable and constrain the acts, and the acts simultaneously change certain structures while also reinforcing others.

21. Kathy Bergen and Greg Burns, "Series of Natural Disasters Force Huge Payouts, Leaving Companies Vulnerable," *Chicago Tribune,* 7 May 1998; "Weather's Price Tag in '98 Tops 1980s' Tab," *Sacramento Bee,* 28 November 1998, sec. A, 3; Michelle Thomson, "Catastrophe Reinsurance Case Study: East Coast Regional Company," presentation made at the Casualty Actuaries in Reinsurance Seminar, 11 June 1996, Chicago, Illinois, JPO; Ted Steinberg, *Acts of God: The Unnatural History of Natural Disaster in America* (New York: Oxford University Press, 2000).

22. The analysis in this paragraph and the next three draws from Nancy Langston, *Troubled Waters: An Environmental History of the Boundaries between Water and Land in the Inland West* (Seattle: University of Washington Press, forthcoming, 2003), and Langdon Winner, *The Whale and the Reactor: A Search for Limits in an Age of High Technology* (Chicago: University of Chicago Press, 1986). In chapter 6, Langston outlines the practice of "pragmatic adaptive management," which assumes the variability of ecological factors such as uneven yearly precipitation, changing wetlands location, and volatile predator and prey populations. The Fish and Wildlife Service officials in Oregon's Malheur Basin, which she studied, sought not to maintain the basin in some set of fixed ecological conditions but rather managed for variability of the entire

system, seeking to balance the interests of flora and fauna as well as competing land users in the region. Winner argues that technologies require—or at least are strongly compatible with—certain political forms. The political implications of the adoption of technologies, he says, usually go largely undebated. He calls for greater public attention to these implications and for the rejection of technologies that are accompanied by undesirable arrangements of power and authority.

23. On the uneven and often unjust distribution of the costs and benefits of natural disasters in Los Angeles and elsewhere in the United States, see Hurley, *Environmental Inequalities,* and Steinberg, *Acts of God.*

Bibliography

ARTICLES, BOOKS, DISSERTATIONS, AND THESES

Adams, Emma H. *To and Fro in Southern California with Sketches of Arizona and New Mexico*. Cincinnati: W.M.B.C. Press, 1887.

Adams, Paul. "Tunnel Blame Drama Begins." *Baltimore Sun,* 27 July 2001, sec. C, 1.

Adams, Richard N. *Paradoxical Harvest: Energy and Explanation in British History, 1870–1914*. Cambridge: Cambridge University Press, 1982.

Albala-Bertrand, J. M. *Political Economy of Large Natural Disasters, with Special Reference to Developing Countries*. Oxford: Clarendon Press, 1993.

"All South Rallies as $3,000,000 Rain Goes." *Los Angeles Examiner,* 23 February 1914, sec. I, 1–2.

Anderson, M. Kat, Michael G. Barbour, and Valerie Whitworth. "A World of Balance and Plenty: Land, Plants, Animals, and Humans in Pre-European California." In *Contested Eden: California before the Gold Rush,* ed. Ramón A. Gutiérrez and Richard J. Orsi, 12–47. Berkeley: University of California Press, 1998.

Angel, Jacob. "Presidentially Declared Major Disaster for Seven Southern California Counties: The January, February, and March 1980 Storms." In *Storms, Floods, and Debris Flows in Southern California and Arizona, 1978 and 1980: Proceedings of a Symposium, September 17–18, 1980,* sponsored by National Research Council, Committee on Natural Disasters, and Environmental Quality Laboratory, California Institute of Technology, 453–56. Washington, D.C.: National Academy Press, 1982.

"Arroyo Dwellers, Ranchers, Flee Raging Flood." *Los Angeles Examiner,* 22 February 1914, sec. I, 1–2.

Bakker, Elna. *An Island Called California: An Ecological Introduction to Its Natural Communities,* 2d ed. Berkeley: University of California Press, 1984.

Barbour, Michael G. "Ecological Fragmentation in the Fifties." In *Uncommon Ground: Toward Reinventing Nature,* ed. William Cronon, 233–55. New York: W. W. Norton & Company, 1995.

Barron, Hal S., and Richard J. Orsi, ed. *Citriculture and Southern California. California History* 74, Special Issue (spring 1995).

Barry, John. *Rising Tide: The Great Mississippi Flood of 1927 and How It Changed America.* New York: Simon and Schuster, 1997.

Bartlett, Dana. *The Better City.* Los Angeles: Neuner Co. Press, 1907.

Baumann, Paul. "Control of Flood Debris in San Gabriel Area." *Civil Engineering* 14 (April 1944): 144.

Baur, John E. *The Health Seekers of Southern California, 1870–1900.* San Marino, Calif.: Huntington Library, 1959.

Bean, Walton, and James J. Rawls. *California: An Interpretive History,* 6th ed. New York: McGraw-Hill, Inc., 1993.

Beck, Ulrich. *Risk Society: Towards a New Modernity.* Translated by Mark Ritter. London: Sage Publications, 1992.

Bergen, Kathy, and Greg Burns. "Series of Natural Disasters Force Huge Payouts, Leaving Companies Vulnerable." *Chicago Tribune,* 7 May 1998.

Bigger, Richard. *Flood Control in Metropolitan Los Angeles.* Berkeley: University of California Press, 1959.

Bishop, Peter. "Controlled Chaos: Two Approaches to Predicting the Future." *Contingencies* 9 (November/December 1997): 14–18.

Bixby, Floyd Suter. "Drainage Doctors: Modern Construction Efficiency in Los Angeles Flood Control Program." *Constructor,* May 1941, 27–29.

Botkin, Daniel B. *Discordant Harmonies: A New Ecology for the Twenty-first Century.* New York: Oxford University Press, 1990.

Braudel, Fernand. *On History.* Translated by Sarah Matthews. Chicago: University of Chicago Press, 1980.

Braun, Stephen. "Hungry for Power in Georgetown." *Los Angeles Times,* 16 June 2001, sec. A, 10.

Brown, William M., Brent D. Taylor, Oded C. Kolker, Wade G. Wells, II, Nancy R. Palmer, Norman H. Brooks, and Robert C. Y. Koh. "Special Inland Studies." In *Sediment Management for Southern California Mountains, Coastal Plains and Shoreline,* Environmental Quality Laboratory, California Institute of Technology, part D. Pasadena: California Institute of Technology, 1981.

Brownlie, William R., and Brent D. Taylor. "Coastal Sediment Delivery by Major Rivers in Southern California." In *Sediment Management for Southern California Mountains, Coastal Plains and Shoreline,* Environmental Quality Laboratory, California Institute of Technology, part C. Pasadena: California Institute of Technology, 1981.

Buckle, Henry Thomas. *Introduction to the History of Civilization in England.* London, 1857–1861. Reprint. New York: Albert and Charles Boni, 1925.

Bugliarello, George. "Biology, Society and Machines." *American Scientist,* May–June 1998, 232.
Burnham, John C. "A Neglected Field: The History of Natural Disasters." *Perspectives* (April 1988): 22–24.
Calvert, Scott. "Plans to Repair Water Main Hit Another Snag." *Baltimore Sun,* 21 July 2001, sec. A, 6.
Carpenter, Ford A. "Flood Studies at Los Angeles." *Monthly Weather Review* 42 (June 1914): 385–91.
Carson, Rachel. *Silent Spring.* Boston: Houghton Mifflin Company, 1962.
Case, Walter H., ed. *Long Beach Blue Book.* Long Beach: Long Beach Press-Telegram Publishing Company, 1942.
Chandler, Jr., Alfred D. *The Visible Hand: The Managerial Revolution in American Business.* Cambridge: Belknap Press of Harvard University, 1977.
Chaplin, B. F. *Pictorial Story of the Southern California Flood, March 3, 1938.* Anaheim, 1938.
"Checking Torrential Floods." *Engineering News-Record,* 1 July 1937, 22–24.
"Chemical in Manhole Explosions Linked to CSX Train Derailment." Associated Press State and Local Wire, 7 September 2001, online access via Lexis-Nexis.
Cleland, Robert Glass. *The Cattle on a Thousand Hills: Southern California, 1850–1880.* San Marino, Calif.: Huntington Library, 1941.
Clements, Frederic E. *Plant Succession: An Analysis of the Development of Vegetation.* Washington, D.C.: Carnegie Institution of Washington, 1916.
Coburn, Judith. "Whose River Is It, Anyway?" *Los Angeles Times Magazine,* 20 November 1994, 19–24, 48–54.
"Community Action in the West: Sierra Madre Creates a Wilderness Park." *Sunset,* May 1967, 34–35.
"Complete Election Returns from City and County." *Los Angeles Times,* 9 November 1934, sec. I, 2.
"Conclusions of E. C. Eaton, Chief Flood Control Engineer." *Municipal League of Los Angeles Bulletin,* 1 February 1930, 6.
Coquery-Vidrovitch, Catherine. "Is L.A. a Model or a Mess?" *American Historical Review* 105 (December 2000): 1683–91.
"The Cost of Dams." *Los Angeles Times,* 10 October 1934, sec. II, 4.
Cottle, Elmer James. *Grandpa Remembers: Life in Montana and California, 1903–1945.* Privately published by Cottle, 1990. Rare Books Collection, Huntington Library, San Marino, California.
Crandell, John. "Rio Porciuncula: A New Perspective on the Former Environs of Los Angeles." *Southern California Quarterly* 81 (fall 1999): 305–14.
Creager, William P. "Possible and Probable Future Floods: Analysis of Data on 'Record Floods' Suggests Need for Comprehensive Program of Research in Storm Probabilities." *Civil Engineering* 9 (November 1939): 669–70.
Cronon, William. *Changes in the Land: Indians, Colonists, and the Ecology of New England.* New York: Hill and Wang, 1983.
———. "Cutting Loose or Running Aground?" *Journal of Historical Geography* 20 (1994): 38–43.
———. *Nature's Metropolis: Chicago and the Great West.* New York: W. W. Norton & Company, 1991.

———. "A Place for Stories: Nature, History, and Narrative." *Journal of American History* 78 (March 1992): 1347–76.
———. "The Trouble with Wilderness; or, Getting Back to the Wrong Nature." In *Uncommon Ground: Toward Reinventing Nature*, ed. William Cronon, 69–90. New York: W. W. Norton & Company, 1995.
———. "The Uses of Environmental History." *Environmental History Review* 17 (fall 1993): 1–22.
———, ed. *Uncommon Ground: Toward Reinventing Nature.* New York: W. W. Norton & Company, 1995.
Crouch, Winston W. *Metropolitan Los Angeles, A Study in Integration: Intergovernmental Relations*, vol. 15. Los Angeles: Haynes Foundation, 1954.
Crouch, Winston W., and Beatrice Dinerman. *Southern California Metropolis: A Study in Development of Government for a Metropolitan Area.* Berkeley: University of California Press, 1963.
Cruse, R. E. "Structures to Control Torrents." *Engineering News-Record*, 8 July 1937, 69–70.
"Dam May Be Redesigned." *Los Angeles Times*, 6 November 1934.
"Dams' Cost Climbing Up." *Los Angeles Times*, 9 October 1934, sec. II, 1.
"Dam Work Suspended." *Los Angeles Times*, 10 November 1934, sec. II, 5.
Davis, Clark. "From Oasis to Metropolis." *Pacific Historical Review* 61 (August 1992): 357–86.
Davis, J. Daniel. "Rare and Unusual Postfire Flood Events Experienced in Los Angeles County During 1978 and 1980." In *Storms, Floods, and Debris Flows in Southern California and Arizona, 1978 and 1980: Proceedings of a Symposium, September 17–18, 1980*, sponsored by National Research Council, Committee on Natural Disasters, and Environmental Quality Laboratory, California Institute of Technology, 243–56. Washington, D.C.: National Academy Press, 1982.
Davis, Mike. "The Case for Letting Malibu Burn." *Environmental History Review* 19 (summer 1995): 1–36.
———. *City of Quartz: Excavating the Future in Los Angeles.* London: Verso, 1990.
———. *Ecology of Fear: Los Angeles and the Imagination of Disaster.* New York: Metropolitan Books, 1998.
———. "How Eden Lost Its Garden: A Political History of the Los Angeles Landscape." In *The City: Los Angeles and Urban Theory at the End of the Twentieth Century*, ed. Allen J. Scott and Edward Soja, 160–85. Berkeley: University of California Press, 1996.
———. "Los Angeles after the Storm: The Dialectic of Ordinary Disaster." *Antipode* 27 (1995): 221–41.
Davis, Tony. "The West's Hottest Question: How to Burn What's Bound to Burn." *High Country News*, 5 June 2000.
Dear, Michael, ed., with J. Dallas Dishman. *From Chicago to L.A.: Making Sense of Urban Theory.* Thousand Oaks, Calif.: Sage Publications, 2002.
Dear, Michael, H. Eric Schockman, and Greg Hise, eds. *Rethinking Los Angeles.* Thousand Oaks, Calif.: Sage Publications, 1996.

deBuys, William. "Los Alamos Fire Offers a Lesson in Humility." *High Country News,* 3 July 2000.
Demerit, David. "Ecology, Objectivity and Critique in Writings on Nature and Human Societies." *Journal of Historical Geography* 20 (1994): 22–37.
Dennet, Daniel. *Darwin's Dangerous Idea: Evolution and the Meanings of Life.* New York: Simon and Schuster, 1995.
Detwiler, Justice B. *Who's Who in California: A Biographical Directory, 1928–1929.* San Francisco: Who's Who Publications, 1929.
Deverell, William F. "The Los Angeles Free Harbor Fight." *California History* 70 (spring 1991): 12–29, 131–35.
Dewar, Heather, Marcia Myers, and Kimberly A. C. Wilson. "Accident Plan Leaves City Unprepared." *Baltimore Sun,* 26 July 2001, sec. A, 1.
DiLeo, Michael, and Eleanor Smith. *Two Californias: The Truth about the Split-State Movement.* Covelo, Calif.: Island Press, 1983.
Donahue, Bill. "The Same River Twice." *Mother Jones,* December 2000, http://www.motherjones.com/mother_jones/NDOO/river.html.
"Dorothy Green: Return Storm Water to Aquifers." *California Coast and Ocean,* summer 1993, 30–31.
Douglas, Ian. *The Urban Environment.* Baltimore: Edward Arnold, 1983.
Drew, Linda G., ed. *Tree-Ring Chronologies of Western America: California and Nevada,* vol. 3. Tucson: Laboratory of Tree-Ring Research, University of Arizona, 1972.
Dvorak, Petula. "Manholes Erupt Again in Georgetown." *Washington Post,* 14 September 2000, sec. B, 3.
Eaton, E. Courtlandt. "Flood and Erosion Control Problems and Their Solution." *Transactions of the American Society of Civil Engineers* 101 (1936): 1302–62.
Eberts, Mike. *Griffith Park: A Centennial History.* Los Angeles: Historical Society of Southern California, 1996.
Edwards, Kenneth L. "Failure of the San Jacinto River Levees Near San Jacinto, California, from the Floods of February 1980." In *Storms, Floods, and Debris Flows in Southern California and Arizona, 1978 and 1980: Proceedings of a Symposium, September 17–18, 1980,* sponsored by National Research Council, Committee on Natural Disasters, and Environmental Quality Laboratory, California Institute of Technology, 347–56. Washington, D.C.: National Academy Press, 1982.
Eliot, Charles W., and Donald F. Griffin. *Waterlines: Key to Development of Metropolitan Los Angeles.* Los Angeles: Haynes Foundation, 1946.
Elkind, Sarah S. *Bay Cities and Water Politics: The Battle for Resources in Boston and Oakland.* Lawrence: University Press of Kansas, 1998.
Engh, Michael E., S.J. "At Home in the Heteropolis: Understanding Postmodern L.A." *American Historical Review* 105 (December 2000): 1676–82.
———. *Frontier Faiths: Church, Temple, and Synagogue in Los Angeles, 1848–1888.* Albuquerque: University of New Mexico Press, 1992.
Engstrom, Wayne N. "Nineteenth-Century Coastal Gales of Southern California." *Geographical Review* 84 (July 1994): 306–15.

Erikson, Kai T. *Everything in Its Path: Destruction of Community in the Buffalo Creek Flood.* New York: Simon and Schuster, 1976.

Fairbanks, Robert B., and Patricia Mooney-Melvin. *Making Sense of the City: Local Government, Civic Culture, and Community Life in Urban America.* Columbus: Ohio State University Press, 2001.

Falick, Abraham Johnson. "Transport Planning in Los Angeles: A Geo-Economic Analysis." Ph.D. diss., University of California, Los Angeles, 1970.

Fall, Edward W. "Regional Geological History." In *Sediment Management for Southern California Mountains, Coastal Plains and Shoreline.* Environmental Quality Laboratory, California Institute of Technology, part A. Pasadena: California Institute of Technology, 1981.

Fehrenbacher, Don E. "The New Political History and the Coming of the Civil War." *Pacific Historical Review* 54 (May 1985): 117–42.

Fesperman, Dan. "With a Rumble, Chaos." *Baltimore Sun,* 21 July 2001, sec. A, 1.

"Fiddling While Parks Burn." *Denver Post,* 14 June 2001, sec. B, 6.

Fiege, Mark. *Irrigated Eden: The Making of an Agricultural Landscape in the American West.* Seattle: University of Washington Press, 1999.

"56,309,600 Tons Water Falls upon Los Angeles." *Los Angeles Examiner,* 1 March 1914, sec. IV, 1.

Fischer, Frank, and John Forester, ed. *The Argumentative Turn in Policy Analysis and Planning.* Durham: Duke University Press, 1993.

Fischer, Frank, and Maarten A. Hajer, eds. *Living with Nature: Environmental Politics as Cultural Discourse.* Oxford: Oxford University Press, 1999.

Fischer, Frank, and Carmen Sirianni, eds. *Critical Studies in Organization and Bureaucracy.* Philadelphia: Temple University Press, 1984.

Fisher, Lewis F. *Crown Jewel of Texas: The Story of San Antonio's River.* San Antonio: Maverick Publishing Company, 1997.

———. *Saving San Antonio: The Precarious Preservation of a Heritage.* Lubbock, Tex.: Texas Tech University Press, 1996.

Fisher, Robert. *Let the People Decide: Neighborhood Organizing in America.* New York: Twayne Publishing, 1994.

Fisher, Robert, and Peter Romanofsky. *Community Organization for Urban Social Change: A Historical Perspective.* Westport, Conn.: Greenwood Press, 1981.

"Five Guests Lose Lives as Flood Ends Gay Party." *Los Angeles Times,* 2 January 1934, sec. I, 2.

Fleischman, John. "Everything You Didn't Know about the L.A. River." *Los Angeles Magazine,* April 1973, 24–26, 78–79.

Fleming, Donald. "The Roots of the New Conservation Movement." *Perspectives in American History* 6 (1971): 3–179.

Fletcher, Russell Holmes, ed. *Who's Who in California: A Biographical Reference Work of Notable Living Men and Women of California,* vol. *1942–1943.* Los Angeles: Who's Who Publications Company, 1941.

"Flood Bonds Are Indorsed [*sic*]." *Los Angeles Times,* 16 April 1924, sec. II, 8.

"Flood Control at Los Angeles." *Engineering News-Record,* 31 January 1935, 177.

"Flood-Control Bonds." *Los Angeles Times,* 30 October 1934, sec. II, 4.
"Flood Insurance: Your First Line of Defense." *USAA Magazine,* August–September 1998, 9–12.
"Floods and the Future." *Los Angeles Times,* 26 February 1914, sec. II, 4.
"Flood Situation in Los Angeles County." *Municipal League Bulletin,* March 1938, 4.
"Flood Waters Real Menace." *Los Angeles Times,* 26 February 1914, sec. II, 1.
Fogelson, Robert M. *The Fragmented Metropolis: Los Angeles, 1850–1930.* Cambridge, Mass.: Harvard University Press, 1967.
"Foothill Areas Periled as Storm Descends on City." *Los Angeles Times,* 18 October 1934, sec. I, 1, 3.
Ford, John Anson. *Thirty Explosive Years in Los Angeles County.* San Marino, Calif.: Huntington Library, 1961.
Fox, Stephen. *John Muir and His Legacy: The American Conservation Movement.* Boston: Little, Brown and Company, 1981.
"Friends of the L.A. River: Improve It." *California Coast and Ocean,* summer 1993, 26–27.
Friends of the San Gabriel River web site. "Job Announcement." http://www.sangabrielriver.org, accessed on 15 May 2002.
Fuentes, Carlos. *A New Time for Mexico.* Translated by Marina Gutman Castañeda and Carlos Fuentes. Berkeley: University of California Press, 1994.
Galambos, Louis, and Joseph Pratt. *The Rise of the Corporate Commonwealth: U.S. Business and Public Policy in the Twentieth Century.* New York: Basic Books, 1988.
Geffner, Marcie. "Immerse Yourself in This Primer on Flood Insurance." *Los Angeles Times* home edition, 10 May 1998, sec. K, 1.
Gellman, Irwin F. *The Contender: Richard Nixon, the Congress Years, 1946–1952.* New York: Free Press, 1999.
Giddens, Anthony. *The Constitution of Society: Outline of the Theory of Structuration.* Berkeley: University of California Press, 1984.
Gleick, James. *Chaos: Making a New Science.* New York: Penguin Books, 1987.
Golley, Frank Benjamin. *A History of the Ecosystem Concept in Ecology: More Than the Sum of the Parts.* New Haven: Yale University Press, 1993.
Goodstein Phil. *Denver in Our Time,* vol. 1: *Big Money in the Big City.* Denver: New Social Publications, 1999.
Gottlieb, Robert. *Forcing the Spring: The Transformation of the American Environmental Movement.* Washington, D.C.: Island Press, 1993.
———. "Rediscovering the River: A Community-Environment Coalition Secures Urban Parkland for L.A." *Orion Afield,* spring 2002, http://www.oriononline.org/pages/oa/02–2oa/LARiver.html, accessed on 22 May 2002.
Gould, Stephen Jay. "Darwinian Fundamentalism." *New York Review of Books,* 12 June 1997, 34–37.
———. "Evolution: The Pleasures of Pluralism." *New York Review of Books,* 26 June 1997, 47–52.
———. *Wonderful Life: The Burgess Shale and the Nature of History.* New York: W. W. Norton & Company, 1989.

"Graft or More Incompetence?" *Municipal League of Los Angeles Bulletin,* 23 October 1926, 7.
Graham, Sara. "Flooding and Your Mortgage." *Los Angeles Times,* 10 February 1997.
Griswold del Castillo, Richard. *The Los Angeles Barrio, 1850–1890: A Social History.* Berkeley: University of California Press, 1979.
Guinn, James Miller. "Exceptional Years: A History of California Flood and Drought." *Publications of the Historical Society of Southern California* 1 (1890): 33–39.
———. *A History of California and an Extended History of Its Southern Coast Counties Also Containing Biographies of Well-Known Citizens of the Past and Present.* Los Angeles: Historic Record Company, 1907.
———. "How California Escaped State Division." *Publications of the Historical Society of Southern California* 6 (1905): 223–32.
Gumprecht, Blake. "51 Miles of Concrete: The Exploitation and Transformation of the Los Angeles River." Master's thesis, California State University, Los Angeles, 1995.
———. *The Los Angeles River: Its Life, Death, and Possible Re-birth.* Baltimore: Johns Hopkins University Press, 1999.
Gutiérrez, David G. *Walls and Mirrors: Mexican Americans, Mexican Immigrants, and the Politics of Ethnicity.* Berkeley: University of California Press, 1995.
Gutiérrez, Ramón A., and Richard J. Orsi. *Contested Eden: California before the Gold Rush.* Berkeley: University of California Press, 1998.
Hackel, Steven W. "Land, Labor, and Production: The Colonial Economy of Spanish and Mexican California." In *Contested Eden: California before the Gold Rush,* ed. Ramón A. Gutiérrez and Richard J. Orsi, 111–46. Berkeley: University of California Press, 1998.
"Halting the Dam." *Los Angeles Times,* 18 October 1934, sec. II, 4.
"Halt on Dam Job Sought." *Los Angeles Times,* 17 October 1934, sec. II, 1.
Harding, S. T. *Water in California.* Palo Alto: N-P Publications, 1960.
Haring, Bruce. "Power Outages Give PC Owners an Unpleasant Jolt." *USA Today,* 19 August 1996, sect. D, 3.
Harvard School of Public Health. "Mosquito Borne Viruses." http://www.hsph.harvard.edu/mosquito/facts.html, accessed on 17 February 2003.
Hays, Samuel P. *Beauty, Health, and Permanence: Environmental Politics in the United Sates, 1955–1985.* New York: Cambridge University Press, 1987.
———. *Conservation and the Gospel of Efficiency: The Progressive Conservation Movement, 1890–1920.* Cambridge, Mass.: Harvard University Press, 1959. Reprint, New York: Atheneum, 1969.
Hedger, Harold E. "A Report upon a Combined Flood Control and Water Conservation System for the San Gabriel River Los Angeles County, California." B.S. thesis, University of California, Berkeley, May 1924.
Henderson, George C. "The Forest Fire War in California." *American Forests and Forest Life* 30 (September 1924): 531–34.
Hewitt, Kenneth. "The Idea of Calamity in a Technocratic Age." In *Interpreta-*

tions of Calamity from the Viewpoint of Human Ecology, ed. Kenneth Hewitt, 3–32. Boston: Allen & Unwin, 1983.

———. *Regions of Risk: A Geographical Introduction to Disasters.* Essex, England: Addison Wesley Longman, 1997.

"Highest Dam in World Is Under Way Near Azusa." *Los Angeles Times,* 31 October 1926, sec. V, 1.

Hill, Robert T. *Southern California Geology and Los Angeles Earthquakes.* Los Angeles: Southern California Academy of Sciences, 1928.

Hise, Greg. *Magnetic Los Angeles: Planning the Twentieth-Century Metropolis.* Baltimore: Johns Hopkins University Press, 1997.

Hise, Greg, and William Deverell, eds. *Eden by Design: The 1930 Olmsted-Bartholomew Plan for the Los Angeles Region.* Berkeley: University of California Press, 2000.

Hittell, Theodore H. *History of California,* vol. 3. San Francisco: N. J. Stone & Company, 1898.

Hoffman, Abraham. *Vision or Villainy: Origins of the Owens Valley–Los Angeles Water Controversy.* College Station, Tex.: Texas A&M Press, 1981.

Hotz, Robert Lee, and Frank Clifford. "A Glitch in the System." *Los Angeles Times,* 14 August 1996, sec. A, 1.

Hoult, Thomas Ford. "The Whittier Narrows Dam: A Study in Community Competition and Conflict." M.A. thesis, Whittier College, 1948.

Hughes, Thomas P. *Networks of Power: Electrification in Western Society, 1880–1930.* Baltimore: Johns Hopkins University Press, 1983.

Hundley, Norris, Jr. *The Great Thirst: Californians and Water, 1770s–1990s.* Berkeley: University of California Press, 1992.

———. *Water and the West: The Colorado River Compact and the Politics of Water in the American West.* Berkeley: University of California Press, 1975.

Hurley, Andrew, ed. *Common Fields: An Environmental History of St. Louis.* St. Louis: Missouri Historical Society, 1997.

———. *Environmental Inequalities: Class, Race, and Industrial Pollution in Gary, Indiana, 1945–1980.* Durham: University of North Carolina Press, 1995.

"Indorses [*sic*] Flood Control Bill." *Los Angeles Times,* 10 June 1915, sec. I, 12.

Jackson, Donald C. *Building the Ultimate Dam: John S. Eastwood and the Control of Water in the West.* Lawrence: University Press of Kansas, 1995.

Jones, E. L. *The European Miracle: Environment, Economics and Geopolitics in the History of Europe and Asia.* Cambridge: Cambridge University Press, 1987.

Joseph, Gilbert M., and Daniel Nugent, eds. *Everyday Forms of State Formation: Revolution and the Negotiation of Rule in Modern Mexico.* Durham: Duke University Press, 1994.

Josselyn, Michael, and Sarah Chamberlain. "History: The Way It Was." *California Coast and Ocean,* summer 1993, 20–23.

Kahn, Judd. *Imperial San Francisco: Politics and Planning in an American City, 1897–1906.* Lincoln: University of Nebraska Press, 1979.

Kahrl, William L. *Water and Power: The Conflict over Los Angeles's Water Supply in the Owens Valley.* Berkeley: University of California Press, 1982.

Keller, David R., and Frank B. Golley, eds. *The Philosophy of Ecology: From Science to Synthesis.* Athens, Ga.: University of Georgia Press, 2000.

Kelley, Robert. *Battling the Inland Sea: American Political Culture, Public Policy, and the Sacramento Valley, 1850–1986.* Berkeley: University of California Press, 1989.

———. "Taming the Sacramento: Hamiltonianism in Action." *Pacific Historical Review* 34 (February 1965): 21–49.

Kiehl, Stephen. "Fact-Finding Phase of Probe into Train Tunnel Fire Nears End." *Baltimore Sun,* 10 December 2002, sec. B, 10.

Kinney, Abbot. *Forest and Water.* Los Angeles: Post Publishing Company, 1900.

Kline, Benjamin. *First along the River: A Brief History of the U.S. Environmental Movement.* San Francisco: Acada Books, 2000.

"Knabe Announces Elimination of Flood Insurance for Lower Los Angeles River Area Homeowners." Press release from office of Supervisor Don Knabe. 11 January 2002. See http://ladpw.org/wmd/LACDA?press_release.pdf, accessed on 3 May 2002.

Knudson, Tom, and Nancy Vogel. "Superfloods Threaten West." *Sacramento Bee,* 23 November 1997.

Kraebel, Charles J. "The La Crescenta Flood: Real Origin of California's New Year Catastrophe Traced to Mountain Slopes Recently Swept by Fire." *American Forests* 40 (June 1934): 251–54, 286–87.

Kretzmann, Stephen P. "A House Built upon the Sand: Race, Class, Gender, and the Galveston Hurricane of 1900." Ph.D. diss., University of Wisconsin–Madison, 1995.

Kroll, Christopher, and "R G." "What River? Changing Views of the River." *California Coast and Ocean,* summer 1993, 32–34.

"Lack of Funds Prevents Curb on Flood Disasters." *Los Angeles Times,* 19 October 1934, sec. I, 9.

"L.A. Dams Prevented Disaster." *Los Angeles Herald-Examiner,* 28 November 1965.

"L.A. Flood Control District Reports on March 2 Flood." *Municipal League Bulletin,* August 1938, 7.

Lakshmanan, Indira A. R. "China's Pride and Pain: Three Gorges Dam Is a Monument to Some, an Endless Fountain of Controversy to Others." *Boston Globe,* 6 November 2000, sec. A, 1.

Langston, Nancy. *Forest Dreams, Forest Nightmares: The Paradox of Old Growth in the Inland West.* Seattle: University of Washington Press, 1995.

———. *Troubled Waters: An Environmental History of the Boundaries between Water and Land in the Inland West.* Seattle: University of Washington Press, 2003.

———. *Where Land and Water Meet: A Western Landscape Transformed.* Seattle: University of Washington Press, 2003.

"League Asks Grand Jury Probe." *Municipal League of Los Angeles Bulletin,* 30 April 1926, 1.

Lee, Elsworth. "The Creek That Never Touches Land." *Westways,* September 1966, 28–29.

Leonard, Jack. "Flood Insurance Rate Hike Is Canceled." *Los Angeles Times* home edition, 3 July 1998, sec. B, 5.

———. "A Look Ahead." *Los Angeles Times* home edition, 29 June 1998, sec. B, 1.

Leopold, Aldo. *A Sand County Almanac with Other Essays on Conservation from Round River.* New York: Oxford University Press, 1966.

Liebhardt, Barbara. "Interpretation and Causal Analysis: Theories in Environmental History." *Environmental Review* 12 (spring 1988): 23–36.

Lockmann, Ronald F. *Guarding the Forests of Southern California: Evolving Attitudes toward Conservation of Watershed, Woodlands, and Wilderness.* Glendale, Calif.: Arthur H. Clark Company, 1981.

Logan, Michael F. *The Lessening Stream: An Environmental History of the Santa Cruz River.* Tucson: University of Arizona Press, 2002.

"Los Angeles Harbor Is Filled with Debris." *Los Angeles Times,* 22 February 1914, sec. I, 11.

"The Los Angeles River Is: A Big Joke, a Killer in Concrete, a 50-Mile-Long Ditch, All of the Above." *Aqueduct,* spring 1978, 4–9.

"Los Angeles River Long Vexes City with Tricks: Wild Career of Flood Ravages and Shifting Channel Told in Review." *Los Angeles Times,* 11 May 1924.

Luckin, Bill. "Accidents, Disasters and Cities." *Urban History* 20 (October 1993): 5–18.

Lukens, Theodore Parker. *Forestry in Relation to City Building.* Pasadena: Throop College of Technology, 1915.

———. *Pasadena, California, Illustrated and Described, Showing Its Advantages as a Place for Desirable Homes.* Pasadena, 1886.

Lytle, Marcus Z. "Vanishing Rivers." *High Country,* summer 1978, 37–42.

Maass, Arthur. *Muddy Waters: The Army Engineers and the Nation's Rivers.* Cambridge, Mass.: Harvard University Press, 1951.

MacAdams, Lewis. "Restoring the Los Angeles River: A Forty-Year Art Project." *Whole Earth Review,* spring 1995, 63–67.

McDow, Roberta M. "California's Hundred Year Debate: To Divide or Not to Divide." *Pacific Historian* 10 (fall 1966): 22–33.

———. "State Separation Schemes, 1907–1921." *California Historical Society Quarterly* 49 (March 1970): 39–46.

McEvoy, Arthur F. "Conservation and the Environment." In *Encyclopedia of the United States in the Twentieth Century,* vol. III, ed. Stanley I. Kutler, 1357–82. New York: Charles Scribner's Sons, 1996.

———. *Fisherman's Problem: Ecology and Law in the California Fisheries, 1850–1980.* Cambridge: Cambridge University Press, 1986.

———. "Toward an Interactive Theory of Nature and Culture: Ecology, Production, and Cognition in the California Fishing Industry." *Environmental Review* 11 (winter 1987): 289–305.

———. "Working Environments: An Ecological Approach to Industrial Health and Safety." *Technology and Culture* 36 (April 1995): 145–72.

McPhee, John. *The Control of Nature.* New York: Farrar, Straus, Giroux, 1989.

McShane, Clay. *Down the Asphalt Path: The Automobile and the American City.* New York: Columbia University Press, 1994.

McWilliams, Carey. *Southern California: An Island on the Land.* New York: Duell, Sloan & Pearce, 1946. Reprint, Salt Lake City: Peregrine Smith Books, 1994.

Malnic, Eric, and Henry Chu. "L.A. Region Faces Major Flood Risk, U.S. Warns." *Los Angeles Times,* 25 October 1997, http://www.latimes.com, accessed on 22 February 1999.

"Mammoth Undertaking Fathered by County." *Los Angeles Times,* 2 July 1914.

Matthias, N. A. "The Los Angeles Flood Control Project." *Military Engineer,* September 1941, 382–88.

May, Robert M. "Biological Populations with Nonoverlapping Generations: Stable Points, Stable Cycles, and Chaos." *Science* 186 (15 November 1974): 645–47.

Melosi, Martin V. *Effluent America: Cities, Industry, Energy, and the Environment.* Pittsburgh: University of Pittsburgh Press, 2001.

———. *Garbage in the Cities: Refuse, Reform, and the Environment, 1880–1980.* College Station, Tex.: Texas A&M University Press, 1981.

———. *The Sanitary City: Urban Infrastructure in American from Colonial Times to the Present.* Baltimore: Johns Hopkins University Press, 2000.

Merchant, Carolyn. *The Death of Nature: Women, Ecology, and the Scientific Revolution.* San Francisco: HarperSanFrancisco, 1980.

———. *Ecological Revolutions: Nature, Gender, and Science in New England.* Chapel Hill: University of North Carolina Press, 1989.

Miller, Char. "Flood of Memories." *Texas Observer,* 27 September 1996, 16–17.

Miller, Jeff. "Denver's New Front Yard—Good Times Down by the Riverside." *Boston Globe,* 3 June 2001, sec. F, 16.

Minnich, Richard. *The Biogeography of Fire in the San Bernardino Mountains: A Historical Study.* University of California Publications, Geography, vol. 28. Berkeley: University of California Press, 1988.

Mooney-Melvin, Patricia. "Before the Neighborhood Organization Revolution: Cincinnati's Neighborhood Improvement Associations, 1890–1940." In *Making Sense of the City: Local Government, Civic Culture, and Community Life in Urban America,* ed. Robert B. Fairbanks and Patricia Mooney-Melvin, 95–118. Columbus: Ohio State University Press, 2001.

Moore, John M., ed. *Moore's Who Is Who in California.* Los Angeles: John M. Moore, Publisher, 1958.

"Mopping Up." *Los Angeles Times,* 4 March 1938, sec. II, 4.

Morrison, Patt; photographs by Mark Lamonica. *Río L.A.: Tales from the Los Angeles River.* Santa Monica, Calif.: Angel City Press, 2001.

"Mother Nature's 'Great Mischief.'" *Chicago Tribune,* 27 August 1998, sec. I, 22.

Mount, Jeffrey F. *California Rivers and Streams: The Conflict between Fluvial Process and Land Use.* Berkeley: University of California Press, 1995.

"Move Spurred for Added San Gabriel River Outlet." *Los Angeles Times,* 8 March 1938, sec. I, 3.

Mumford, Lewis. *Technics and Civilization.* New York: Harcourt, Brace and Company, 1934.

Nash, Roderick. *Wilderness and the American Mind.* New Haven: Yale University Press, 1982.

National Research Council (NRC), Committee on Natural Disasters, and Environmental Quality Laboratory, California Institute of Technology. *Storms, Floods, and Debris Flows in Southern California and Arizona, 1978 and 1980: Overview and Summary of a Symposium, September 17–18, 1980.* Washington, D.C.: National Academy Press, 1982.

———. *Storms, Floods, and Debris Flows in Southern California and Arizona, 1978 and 1980: Proceedings of Symposium, September 17–18, 1980.* Washington, D.C.: National Academy Press, 1982.

"New City Charter Wins; Police Bonds Approved." *Los Angeles Times,* 7 May 1924, sec. I, 1.

"New Cornfield Proposal Includes Canal and School." *Los Angeles Times,* 20 April 2001.

Newton, C. T., and Harold E. Hedger. "Los Angeles County Flood Control and Water Conservation." *Journal of the Waterways and Harbors Division, Proceedings of the American Society of Civil Engineers* 85 ww2 (June 1959): 81–97.

Nixon, Edgar B., ed. *Franklin D. Roosevelt and Conservation, 1911–1945.* Washington, D.C.: Government Printing Office, 1957.

Nordhoff, Charles. *California: For Health, Pleasure, and Residence: A Book for Travellers and Settlers.* New York: Harper & Brothers, Publishers, 1873.

"No Unemployed." *Los Angeles Times,* 25 February 1914, sec. II, 4.

Nunis, Doyce B., Jr., ed. *The St. Francis Dam Disaster Revisited.* Los Angeles and Ventura: Historical Society of Southern California and Ventura County Museum of History and Art, 1995.

Oberbeck, Grace Johnson. *History of La Crescenta–La Cañada Valleys: A Story of Beginnings Put into Writing.* Montrose, Calif.: Ledger, 1938.

Olmos, Robert. "The River: The Mighty 'River of Angels' as It Looked to a Boy in the '30s." *Los Angeles Magazine,* August 1961, 18.

Olmsted, Frank. "Theory and Practice in Flood Control." *Municipal League of Los Angeles Bulletin,* 1 February 1930, 6.

Olmsted Brothers and Bartholomew and Associates. *Parks, Playgrounds and Beaches for the Los Angeles Region.* Los Angeles: Olmsted Brothers and Bartholomew and Associates, 1930.

Orsi, Richard J. "Selling the Golden State: A Study of Boosterism in Nineteenth-Century California." Ph.D. diss., University of Wisconsin–Madison, 1973.

Osmundson, Theodore. "How to Control the Flood Controllers." *Cry California,* summer 1970, 30–38.

Outland, Charles F. *Man-Made Disaster: The Story of St. Francis Dam.* Glendale, Calif.: Arthur Clark Company, 1977.

Perrow, Charles. *Normal Accidents: Living with High-Risk Technologies.* New York: Basic Books, 1984.

Pertman, Adam. "Power Outage Strikes 7 Western States." *Boston Globe,* 11 August 1996, sec. B, 2.

"Peter Goodwin: Consider These Ideas." *California Coast and Ocean,* summer 1993, 28–29.

Pfatteicher, Sarah K. A. "Death by Design: Ethics, Responsibility, and Failure in the American Civil Engineering Community, 1852–1986." Ph.D. diss., University of Wisconsin–Madison, 1996.

Pisani, Donald J. *From the Family Farm to Agribusiness: The Irrigation Crusade in California and the West, 1850–1931.* Berkeley: University of California Press, 1984.

Pitt, Leonard, and Dale Pitt. *Los Angeles A to Z: An Encyclopedia of the City and County.* Berkeley: University of California Press, 1997.

"Pocket Veto." *Los Angeles Times,* 13 June 1915, sec. I, 3.

Post, John D. *The Last Great Subsistence Crisis in the Western World.* Baltimore: Johns Hopkins University Press, 1977.

Prendergast, J. J. "Upstream Flood Control." *California,* December 1939, 10–11.

"Preparing for Failure by Failing to Prepare." *Baltimore Sun,* 27 July 2001, sec. A, 22.

Press Reference Library (Western Edition). *Notables of the West,* vol. 1. New York: International News Service, 1913.

Preston, William. "Serpent in the Garden: Environmental Change in Colonial California." In *Contested Eden: California before the Gold Rush,* ed. Ramón A. Gutiérrez and Richard J. Orsi, 260–98. Berkeley: University of California Press, 1998.

Price, Jennifer. "Paradise Reclaimed: A Field Guide to the L.A. River." *L.A. Weekly,* 10–16 August 2001, 23–39.

Prigogine, Ilya, and Isabelle Stengers. *Order out of Chaos: Man's New Dialogue with Nature.* Boulder, Colo.: New Science Library, 1984.

Pulido, Laura. "Multiracial Organizing Among Environmental Justice Activists in Los Angeles." In *Rethinking Los Angeles,* ed. Michael J. Dear, H. Eric Schockman, and Greg Hise, 171–188. Thousand Oaks, Calif.: Sage Publications, 1996.

Pyke, Charles. "Return Periods of 1977–80 Precipitation in Southern California and Arizona." In *Storms, Floods, and Debris Flows in Southern California and Arizona, 1978 and 1980: Proceedings of a Symposium, September 17–18, 1980,* sponsored by National Research Council, Committee on Natural Disasters, and Environmental Quality Laboratory, California Institute of Technology, 77–86. Washington, D.C.: National Academy Press, 1982.

Pyne, Stephen J. *Fire in America: A Cultural History of Wildland and Rural Fire.* Princeton: Princeton University Press, 1982.

Queenan, Charles F. *The Port of Los Angeles: From Wilderness to World Port.* Los Angeles: Los Angeles Harbor Department, 1983.

"Quick Action on Flood Problem Is Assured." *Los Angeles Times,* 28 February 1914, sec. II, 1.

Rabin, Jeffrey L. "Supervisors OK Plan to Raise Walls of L.A. River." *Los Angeles Times,* 4 September 1996, sec. B, 1, 6.

"The Rain." *Los Angeles Times,* 19 October 1934, sec. II, 4.

Rawls, James J. "The California Mission as Symbol and Myth." *California History* 71 (fall 1992): 342–61, 449–51.
"Reagan and the Board of Supervisors." *Municipal League of Los Angeles Bulletin,* July 1925, 1.
"Reagan's Dam Bonds—Viewing the Remains." *Municipal League of Los Angeles Bulletin,* 25 November 1926, 4–5.
"Reagan Unfair to San Gabriel Farmers." *Municipal League of Los Angeles Bulletin,* 20 August 1926, 1.
Recreating the River: A Prescription for Los Angeles. Santa Monica: Southern California Institute of Architecture, [1990?].
Reisner, Marc. *Cadillac Desert: The American West and Its Disappearing Water.* New York: Penguin Books, 1986.
"Report Rejects Disciplining of Parks Officials in Wildfire." *New York Times,* 13 June 2001, sec. A, 28.
Ricci, James. "Snapshots from the Center of the Universe; A Park with No Name (Yet) but Plenty of History." *Los Angeles Times Magazine,* 15 July 2001, 5.
Rice, Richard B., William A. Bullough, and Richard J. Orsi. *The Elusive Eden: A New History of California,* 3d ed. New York: McGraw-Hill, 2002.
"River Flood Analyzed." *Los Angeles Times,* 12 March 1938, sec. I, 5.
Robbins, George W., and L. Deming Tilton, eds. *Los Angeles: Preface to a Master Plan.* Los Angeles: Pacific Southwest Academy, 1941.
Robinson, Charles Mulford. *The City Beautiful: Report to the Municipal Art Commission.* Los Angeles, 1909.
Robinson, William Wilcox. *The Forests and the People: The Story of the Los Angeles National Forest.* Los Angeles: Title Insurance and Trust Company, 1946.
Rolle, Andrew. *Los Angeles: From Pueblo to City of the Future.* Sacramento: MTL, 1995.
Rome, Adam. *The Bulldozer in the Countryside: Suburban Sprawl and the Rise of American Environmentalism.* Cambridge: Cambridge University Press, 2001.
Roraback, Dick. "The L.A. River Practices Own Trickle-Down Theory." *Los Angeles Times,* 27 October 1985, sec. VI, 1, 6–7.
———. "Up a Lazy River, Seeking the Source: Your Explorer Follows in Footsteps of Gaspar de Portolá." *Los Angeles Times,* 20 October 1985, sec. I, 1–2.
Rosen, Christine Meisner. *The Limits of Power: Great Fires and the Process of City Growth in America.* Cambridge: Cambridge University Press, 1986.
Rosenberg, Charles E. *The Cholera Years: The United States in 1832, 1849 and 1866.* Chicago: University of Chicago Press, 1962.
Rothman, Hal K. *The Greening of a Nation? Environmentalism in the United States since 1945.* Fort Worth: Harcourt Brace, 1998.
"A Round Table: Environmental History." *Journal of American History* 76 (March 1990): 1087–147.
Rowntree, Lester B. "A Crop-Based Rainfall Chronology for Pre-Instrumental Record Southern California." *Climate Change* 7 (1985): 327–41.

"St. Francis Dam Catastrophe—A Great Foundation Failure." *Engineering News-Record,* 22 March 1928, 466–69.

"St. Francis Dam of Los Angeles Water-Supply System Fails under Full Head." *Engineering News-Record,* 15 March 1928, 456.

Sánchez, George J. *Becoming Mexican American: Ethnicity, Culture and Identity in Chicano Los Angeles, 1900–1945.* New York: Oxford University Press, 1993.

Sauer, Carl O. *Land and Life: A Selection from the Writings of Carl Ortwin Sauer.* Edited by John Leighly. Berkeley: University of California Press, 1963.

Sawislak, Karen. *Smoldering City: Chicagoans and the Great Fire, 1871–1874.* Chicago: University of Chicago Press, 1995.

Schatzberg, Eric. "Ideology and Technical Choice: The Decline of the Wooden Airplane in the United States, 1920–1945." *Technology and Culture* 35 (January 1994): 34–69.

Schmidt, Alfred. *The Concept of Nature in Marx.* London: NLB, 1973.

Schneider, Robert A. "The Postmodern City from an Early Modern Perspective." *American Historical Review* 105 (December 2000): 1668–75.

Sciandrone, Joe, Ted Albrecht, Jr., Richard Davidson, Jacob Douma, Dave Hammer, Charles Hooppaw, and Al Robles, Jr. "Levee Failure and Distress, San Jacinto River Levee and Bautista Creek Channel, Riverside County, Santa Ana River Basin, California." In *Storms, Floods, and Debris Flows in Southern California and Arizona, 1978 and 1980: Proceedings of a Symposium, September 17–18, 1980,* sponsored by National Research Council, Committee on Natural Disasters, and Environmental Quality Laboratory, California Institute of Technology, 357–86. Washington, D.C.: National Academy Press, 1982.

Scott, Allen J., and Edward W. Soja, eds. *The City: Los Angeles and Urban Theory at the End of the Twentieth Century.* Berkeley: University of California Press, 1996.

Scott, Joan Wallach. *Gender and the Politics of History.* New York: Columbia University Press, 1988.

"A Severe Indictment." *Municipal League of Los Angeles Bulletin,* 30 April 1926, 4–5.

Shoemaker, Joe. *Returning the Platte to the People: A Story of A Unique Committee, the Platte River Development Committee.* Westminster, Colo.: Greenway Foundation, 1981.

Shuirman, Gerard, and James E. Slosson. *Forensic Engineering: Environmental Case Histories for Civil Engineers and Geologists.* San Diego: Academic Press, 1992.

Shuit, Douglas P. "1,200 Protest Costly Change in Flood Maps." *Los Angeles Times* home edition, 13 March 1998, sec. B, 1.

"Sierra Club Challenges County Flood Control District on Project." *Los Angeles Times,* 30 November 1975, sec. II, 1.

Simberloff, Daniel. "A Succession of Paradigms in Ecology: Essentialism to Materialism and Probabilism." In *The Philosophy of Ecology: From Science to Synthesis,* ed. David R. Keller and Frank B. Golley. Athens, Ga.: University of Georgia Press, 2000.

Skowronek, Stephen. *Building a New American State: The Expansion of the National Administrative Capacities, 1877–1920.* Cambridge: Cambridge University Press, 1982.

Slosson, James E., and James P. Krohn. "Southern California Landslides of 1978 and 1980." In *Storms, Floods, and Debris Flows in Southern California and Arizona, 1978 and 1980: Proceedings of a Symposium, September 17–18, 1980,* sponsored by National Research Council, Committee on Natural Disasters, and Environmental Quality Laboratory, California Institute of Technology, 291–320. Washington, D.C.: National Academy Press, 1982.

Sly, Liz. "Three Gorges Dam: Great Leap of Faith." *Chicago Tribune,* 20 December 1996, sec. 1, 1, 11.

"Small Losses Prove Value of Dam System." *Los Angeles Times,* 4 March 1938, sec. I, 7.

Smith, Jack. "Driving down the River." *Westways,* July 1974, 38–41, 66–67.

Smith, Michael L. *Pacific Visions: California Scientists and the Environment, 1850–1915.* New Haven: Yale University Press, 1987.

Smith, Neil. *Uneven Development: Nature, Capital and the Production of Space.* Oxford, England: Basil Blackwell, 1984.

Snyder, Gary. "Night Song of the Los Angeles Basin." *Whole Earth Review,* spring 1995, 62.

"Something Doing Every Minute." *Los Angeles Times,* 26 February 1914, sec. II, 4.

"Southland Starts Rebuilding." *Los Angeles Times,* 5 March 1938, sec. I, 1, 4.

"South's Most Severe Storm Claims 5; Loss Thousands." *Los Angeles Examiner,* 21 February 1914, sec. I, 1–2.

Spirn, Anne Whiston. *The Granite Garden: Urban Nature and Human Design.* New York: Basic Books, 1984.

Spotts, Peter N. "It Could Get Hotter in Japan Thanks to Three Gorges Dam." *Christian Science Monitor,* 12 April 2001, Features section, 14.

"Stalking the Common Mosquito—and a Deadly Virus." National Public Radio, http://www.npr.org/programs/morning/features/2001/jul/010713.mosquito.html, accessed on 17 February 2003.

Starr, Kevin. *Inventing the Dream: California through the Progressive Era.* New York: Oxford University Press, 1985.

———. *Material Dreams: Southern California through the 1920s.* New York: Oxford University Press, 1990.

Stein, George. "Study Widens Risk Zone of Hypothetical Major Flood." *Los Angeles Times,* 15 September 1987, sec. II, 1, 6.

Steinberg, Ted. *Acts of God: The Unnatural History of Natural Disaster in America.* New York: Oxford University Press, 2000.

———. "Do-It-Yourself Deathscape: The Unnatural History of Natural Disaster in South Florida." *Environmental History* 2 (October 1997): 414–38.

Stewart, Jill. "Forget It, Jake. It's Chinatown." New Times Los Angeles Online, 30 March–5 April 2000, http://www.newtimesla.com/issues/current/stewart.html, accessed on 31 March 2000.

Stoll, Steven. *The Fruits of Natural Advantage: Making the Industrial Countryside in California.* Berkeley: University of California Press, 1998.

"Storm Object Lesson: City Must Prepare for Future, Says Mayor." *Los Angeles Examiner,* 22 February 1914, sec. I, 3.

"Storm Spawned by Shift in Weather Thousands of Miles Away." *Los Angeles Times,* 11 February 1978, sec. I, 2, 26.

Sumner, William Graham. *Folkways: A Study of the Sociological Importance of Usages, Manners, Customs, Mores, and Morals.* Originally published 1906. Reprint Boston: Ginn & Company, 1940.

"Symposium on Flood Control of Los Angeles County." *Municipal League of Los Angeles Bulletin,* 1 February 1930.

Tait, Robert. "Georgetown's Cast Iron Excuse for a Crisis." *Scotsman,* 16 June 2001, 14.

Tarr, Joel A. *The Search for the Ultimate Sink: Urban Pollution in Historical Perspective.* Akron: University of Akron Press, 1996.

Tarr, Joel A., and Gabriel Dupuy, eds. *Technology and the Rise of the Networked City in Europe and America.* Philadelphia: Temple University Press, 1988.

Taylor, Brent D. "Inland Sediment Movements by Natural Processes." In *Sediment Management for Southern California Mountains, Coastal Plains and Shoreline.* Environmental Quality Laboratory, California Institute of Technology, part B. Pasadena: California Institute of Technology, 1981.

Taylor, Dolores B. "Flood Flows in Major Streams in Ventura County." In *Storms, Floods, and Debris Flows in Southern California and Arizona, 1978 and 1980: Proceedings of a Symposium, September 17–18, 1980,* sponsored by National Research Council, Committee on Natural Disasters, and Environmental Quality Laboratory, California Institute of Technology, 151–64. Washington, D.C.: National Academy Press, 1982.

Taylor Yard web site. http://tayloryard.org/, accessed on 20 May 2002.

Tettemer, John M. "Closing Comments on Debris and Sediment." In *Storms, Floods, and Debris Flows in Southern California and Arizona, 1978 and 1980: Proceedings of a Symposium, September 17–18, 1980,* sponsored by National Research Council, Committee on Natural Disasters, and Environmental Quality Laboratory, California Institute of Technology, 483–87. Washington, D.C.: National Academy Press, 1982.

———. "Sediment Flow Hazards: Special Hydrologic Events." In *Storms, Floods, and Debris Flows in Southern California and Arizona, 1978 and 1980: Proceedings of a Symposium, September 17–18, 1980,* sponsored by National Research Council, Committee on Natural Disasters, and Environmental Quality Laboratory, California Institute of Technology, 215–28. Washington, D.C.: National Academy Press, 1982.

"Three Million Dollars Price of Safe Rivers." *Los Angeles Times,* 27 February 1914, sec. II, 1.

Tobey, Ronald C. *Saving the Prairies: The Life Cycle of the Founding School of American Plant Ecology, 1895–1955.* Berkeley: University of California Press, 1981.

Tobey, Ronald, and Charles Wetherell. "The Citrus Industry and the Revolution of Corporate Capitalism in Southern California, 1887–1944." *California History* 74 (spring 1991): 6–21, 129–31.

"Torrential Rains Leave Seven Dead; $150,000 Damage." *Los Angeles Times,* 19 October 1934, sec. I, 1, 9.
"To Rush Work on Flood Dam." *Los Angeles Times,* 3 October 1923, sec. II, 2.
"Transcription of Press Conference Regarding the Cerro Grande Fire Held May 18, 2000 in Sante [*sic*] Fe, New Mexico, with Secretary Bruce Babbitt." 26 May 2000, http://www.nps.gov/cerrogrande/transcript.htm, accessed on 27 February 2003.
Turhollow, Anthony F. *A History of the Los Angeles District, U.S. Army Corps of Engineers.* Los Angeles: U.S. Army Engineer District, 1975.
"Twenty-five Montrose Homes Victims in Flood." *Los Angeles Times,* 19 October 1934, sec. I, 9.
"A Twisted Flood Control Program." *Municipal League of Los Angeles Bulletin,* 1 December 1929, 3.
"2 Drowned at Covina in Wake of Big Storm." *Los Angeles Examiner,* 20 February 1914, sec. II, 2.
"Two Ships Held in River Silt, Debris Chokes Up Turning Basin." *Los Angeles Examiner,* 22 February 1914, sec. I, 2.
"Unprecedented Rain Paralyzes Traffic; 1 Dead; City Streets Are Rivers; Hundreds Marooned." *Los Angeles Examiner,* 19 February 1914, sec. I, 1.
"Urge Flood Control Dam." *Los Angeles Times,* 2 October 1923, sec. II, 5.
"Value of Flood Control." *Los Angeles Times,* 26 April 1924, sec. II, 4.
Van Wormer, Stephen R. "A History of Flood Control in the Los Angeles County Drainage Area." *Southern California Quarterly* 73 (1991): 55–94.
"Vast Program Launched as Bonds Are Approved." *Los Angeles Times,* 8 May 1924, sec. II, 1, 2.
Vaughan, Diane. *The Challenger Launch Decision: Risky Technology, Culture, and Deviance at NASA.* Chicago: University of Chicago Press, 1996.
Waldie, D. J. *Holy Land: A Suburban Memoir.* New York: St. Martin's Press, 1996.
———. "The Myth of the L.A. River." *Buzz,* April 1996, 80–85, 115, 118.
Walton, John. *Western Times and Water Wars: State, Culture, and Rebellion in California.* Berkeley: University of California Press, 1992.
Warner, J. J. "Inaugural Address of the President, Col. J. J. Warner." *Publications of the Historical Society of Southern California* 1 (1884): 9–18.
"Water Conservation the Most Vital Question in Southland." *Los Angeles Times Farm and Tractor Magazine,* 4 May 1924, 1, 13.
"Water Main Studied in Train Derailment." *New York Times,* 24 July 2001, www.nytimes.com.
"Weather's Price Tag in '98 Tops 1980s' Tab." *Sacramento Bee,* 28 November 1998, sec. A, 3.
Webb, Wellington E. "Restoring Denver's South Platte River." *Urban Land* 57 (April 1998): 16.
Weiss, Marc A. *The Rise of the Community Builders: The American Real Estate Industry and Urban Land Planning.* New York: Columbia University Press, 1987.
Wellock, Thomas Raymond. *Critical Masses: Opposition to Nuclear Power in California, 1958–1978.* Madison: University of Wisconsin Press, 1998.

Wells, Wade G., II. "The Storms of 1978 and 1980 and Their Effect on Sediment Movement in the Eastern San Gabriel Front." In *Storms, Floods, and Debris Flows in Southern California and Arizona, 1978 and 1980: Proceedings of a Symposium, September 17–18, 1980,* sponsored by National Research Council, Committee on Natural Disasters, and Environmental Quality Laboratory, California Institute of Technology, 229–42. Washington, D.C.: National Academy Press, 1982.

White, Charles R. "Lake Elsinore Flood Disaster of March 1980." In *Storms, Floods, and Debris Flows in Southern California and Arizona, 1978 and 1980: Proceedings of a Symposium, September 17–18, 1980,* sponsored by National Research Council, Committee on Natural Disasters, and Environmental Quality Laboratory, California Institute of Technology, 387–400. Washington, D.C.: National Academy Press, 1982.

White, Richard. "American Environmental History: The Development of a New Historical Field." *Pacific Historical Review* 54 (August 1985): 297–335.

———. *The Organic Machine.* New York: Hill and Wang, 1995.

"Who Delays the San Gabriel Dam?" *Municipal League of Los Angeles Bulletin,* 31 July 1926, 1.

"Whole County Busy Remedying Ravages of Storm; Arroyo Homeless Scorn Alms; Demand Safety." *Los Angeles Examiner,* 24 February 1914, sec. I, 2.

"Why Build in Flood Control Channels?" *Municipal League Bulletin,* April 1938, 4.

Widney, J. P. "A Historical Sketch of the Movement for a Political Separation of the Two Californias, Northern and Southern, under Both the Spanish and American Regimes." *Publications of the Historical Society of Southern California* 1 (1888–1889).

Wiebe, Robert H. *The Search For Order, 1877–1920.* New York: Hill and Wang, 1967.

Willard, Charles Dwight. *The Free Harbor Contest at Los Angeles: An Account of the Long Fight Waged by the People of Southern California to Secure a Harbor Located at a Point Open to Competition.* Los Angeles: Kingsley-Barnes & Neuner Co., 1899.

Williams, Clarence, and David A. Fahrenthold. "Manhole Fire Shuts Down Georgetown." *Washington Post,* 14 July 2001, sec. B, 1.

Williams, James C. *Energy and the Making of Modern California.* Akron: University of Akron Press, 1996.

———. "Engineering California Cities." In *Proceedings of the XVIIIth International ICOHTEC Conference,* ed. Alexandre Herlea, 394–400. San Francisco: San Francisco Press, 1993.

Williams, Joan C. "Critical Legal Studies: The Death of Transcendence and the Rise of the New Langdells." *New York University Law Review* 62 (June 1987): 429–96.

Williams, Michael. *Americans and Their Forests: A Historical Geography.* Cambridge: Cambridge University Press, 1989.

Williams, Philip B. "Rethinking Flood-Control Channel Design." *Civil Engineering* (January 1990): 57–59.

Williams, Prentiss. "Los Angeles River: Overflowing with Controversy." *California Coast and Ocean,* summer 1993, 8–19.
Winner, Langdon. *The Whale and the Reactor: A Search for Limits in an Age of High Technology.* Chicago: University of Chicago Press, 1986.
Wise, L. L. "Is Los Angeles Ready for a Flood?" *Engineering News-Record,* 17 May 1956, 40–50.
Woods, Sean. "Tee Time for Tujunga?" *California Coast & Ocean,* summer 1997.
"Worried L.A. River Managers: Control It." *California Coast and Ocean,* summer 1993, 24–25.
Worster, Donald. *Dust Bowl: The Southern Plains in the 1930s.* Oxford: Oxford University Press, 1979.
———. "The Ecology of Order and Chaos." In *Readings in Ecology,* ed. Stanley I. Dodson et al., 77–89. New York: Oxford University Press, 1999.
———. *Nature's Economy: A History of Ecological Ideas,* 2d ed. New York: Cambridge University Press, 1990.
———. *Rivers of Empire: Water, Aridity, and the Growth of the American West.* New York: Pantheon Books, 1985.
Yoon, Ted Young. "An Emergence of a Riverine Ecology: The Los Angeles River." Master's thesis, University of California, Berkeley, 1992.
Young, Terrence. "Moral Order, Language and the Failure of the 1930 Recreation Plan for Los Angeles County." *Planning Perspectives* 16 (2001): 333–56.
Zielbauer, Edward J. "Challenging Flood Control Problems in Southern California." In *Engineering Geology in Southern California,* ed. Richard Lung and Richard Proctor, 289–301. Los Angeles: Association of Engineering Geologists, 1966. Reprint, 1969.
Zunz, Olivier. *Making America Corporate: 1870–1920.* Chicago: University of Chicago Press, 1990.

GOVERNMENT PUBLICATIONS AND REPORTS

This section of the bibliography contains sources that have relatively wide circulation—items that were distributed to the general public or to multiple governmental agencies. Although only one repository is indicated for each source, they can be found in other repositories as well.

United States

Army Corps of Engineers. "Completed LACDA Project Increases Flood Protection." http://www.spl.usace.army.mil/lacda/lacda.html, accessed on 3 May 2002.
Army Corps of Engineers, Los Angeles District (LAD). "Abstract of Analysis of Design: Whittier Narrows Flood Control Basin." 22 January 1952. Water Resources Center Archives, University of California, Berkeley.
———. "Alternative Proposals for Flood Control and Allied Purposes, South

Fork of the Santa Clara River, Santa Clarita Valley, California." December 1975. Water Resources Center Archives, University of California, Berkeley.

———. "Analysis of Design: Whittier Narrows Flood Control Basin." January 1950. Los Angeles County Department of Public Works Technical Library, Alhambra, California.

———. "Appendix D: Report on Floods of January and February 1969 in Los Angeles County." December 1969. Los Angeles County Department of Public Works Technical Library, Alhambra, California.

———. "Final Supplement to Final Detailed Project Report for Flood Control and Allied Purposes, Including EA and Appendixes, South Fork of the Santa Clara River." January 1985. Water Resources Center Archives, University of California, Berkeley.

———. *Flood Control in the Los Angeles County Drainage Area.* Los Angeles, September 1939. Water Resources Center Archives, University of California, Berkeley.

———. "LACDA Update." September 1987. Public Affairs Office, Los Angeles District, U.S. Army Corps of Engineers, Los Angeles.

———. "Los Angeles County Drainage Area Review: Final Feasibility Study Interim Report and Environmental Impact Statement." December 1991; revised June 1992. U.S. Army Corps of Engineers, Public Affairs Office, Los Angeles, California.

———. "Report on Flood Damage in Southern California from Flood of 15–18 January 1952." 12 February 1952. General Administration Files, box 180, Record Group 77, National Archives, Pacific Region, Laguna Niguel, California.

———. "Report on Flood Damage in Southern California from Floods of January 1954." 12 February 1954. General Administration Files, box 181, Record Group 77, National Archives, Pacific Region, Laguna Niguel, California.

———. "Report on Floods of February and March 1978 in Southern California." November 1978. Los Angeles County Department of Public Works Technical Library, Alhambra, California.

———. "Report on Storm and Flood of 25–27 January 1956 in Southern California." 2 March 1956. General Administration Files, box 181, Record Group 77, National Archives, Pacific Region, Laguna Niguel, California.

———. "Report to Board of Consultants on Whittier Narrows Dam and Appurtenant Structures." June 1949. Water Resources Center Archives, University of California, Berkeley.

———. "Restudy of Whittier Narrows Project and Alternative Plans for Flood Control," by R. C. Hunter. December 1946. Los Angeles County Department of Public Works Technical Library, Alhambra, California.

———. "Santa Clara River and Tributaries, California, Interim Review Report for Flood Control: Newhall, Saugus & Vicinity, Los Angeles County." 15 December 1971. Water Resources Center Archives, University of California, Berkeley.

———. "Sepulveda Basin Recreation Lake Feature Design Memorandum." March 1987. Water Resources Center Archives, University of California, Berkeley.

Department of Agriculture. *Survey Report for the Los Angeles River Watershed.* Washington, D.C.: United States Government Printing Office, 1941. Public Affairs Office, Los Angeles District, U.S. Army Corps of Engineers, Los Angeles.

Department of Agriculture, Field Flood Control Coordinating Committee No. 18. "Volume I of Appendices: Appendices A to M to Accompany Survey Report of Los Angeles River" [ca. 1940]. Box 239, Record Group 77, National Archives, Pacific Region, Laguna Niguel, California.

Department of Commerce, National Weather Service. *The Disastrous Southern California and Central Arizona Floods, Flash Floods, and Mudslides of February 1980: A Report to the Administrator.* Natural Disaster Survey Report, National Weather Service, 81–1. Silver Spring, Md.: U.S. Department of Commerce, National Oceanic and Atmospheric Administration, National Weather Service, 1981.

Department of the Interior, Fish and Wildlife Service. *The Ecology of Riparian Habitats of the Southern California Coast Region: A Community Profile,* by Phyllis M. Faber, Ed Keller, Anne Sands, and Barbara M. Massey. Biological Report, U.S. Fish and Wildlife Service, 85 (7.27). September 1989.

Department of the Interior, National Parks Service, Board of Inquiry. "Cerro Grande Prescribed Fire: Board of Inquiry Final Report." 26 February 2001, http://www.nps.gov/fire/, accessed on 13 June 2001.

Department of the Interior, U.S. Geological Survey (USGS). *Effects of Urbanization on Sedimentation and Floodflows in Coma Creek Basin,* by J. M. Knott. California Open File Report, U.S. Department of the Interior, Geological Survey, Water Resources Division, in cooperation with the County of San Mateo, California. Menlo Park, Calif.: U.S. Geological Survey, 1973.

———. *Flood in La Cañada Valley, California, January 1, 1934,* by Harold C. Troxell and John Q. Peterson. Water Supply Paper, U.S. Geological Survey, 796-C. Washington, D.C.: United States Government Printing Office, 1937.

———. *Floods of February 1980 in Southern California and Central Arizona,* by E. H. Chin, B. N. Aldridge, and R. J. Longfield. Professional Paper, U.S. Geological Survey, 1494. Washington, D.C.: U.S. Government Printing Office, 1991.

———. *Floods of March 1938 in Southern California,* by Harold C. Troxell et al. Water Supply Paper, U.S. Geological Survey, 844. Washington, D.C.: United States Government Printing Office, 1938.

———. *Institutional and Management Aspects, Flood and Drought Functions of the U.S. Army Corps of Engineers,* by David Wingerd and Ming T. Tseng. Water Supply Paper, U.S. Geological Survey, 2375. Washington, D.C.: United States Government Printing Office, 1988–1989.

———. *Maps Showing Historical Flooding in the San Fernando Valley, California, 1934–1956.* Reston, Va.: U.S. Geological Survey, 1981.

———. *Soil Slips, Debris Flows, and Rainstorms in the Santa Monica Mountains and Vicinity, Southern California,* by Russell H. Campbell. Professional Paper, U.S. Geological Survey, 851. Washington, D.C.: U.S. Government Printing Office, 1975.

———. *Southern California Floods of January 1916,* by H. D. McGlashan and

F. C. Ebert. Water Supply Paper, U.S. Geological Survey, 426. Washington, D.C.: United States Government Printing Office, 1918.

———. *Urban Sprawl and Flooding in Southern California,* by S. E. Rantz. Circular, U.S. Geological Survey, 601-B. Washington, D.C.: U.S. Department of the Interior, 1970.

———. *Water Resources of Southern California With Special Reference to the Drought of 1944–51,* by Harold C. Troxell. Water Supply Paper, U.S. Geological Survey, 1366. Washington, D.C.: United States Government Printing Office, 1957.

Department of the Interior, U.S. National Interagency Fire Center, Fire Investigation Team. "Cerro Grande Prescribed Fire, May 4–8, 2000, Investigation Report." 18 May 2000. http://www.nps.gov/cerrogrande/, accessed on 13 June 2001.

Department of the Interior, U.S. Soil Conservation Service. "Work Plan for the Santa Susana Creek Subwatershed of the Los Angeles River Flood Prevention Project, Los Angeles County, California." April 1960. Water Resources Center Archives, University of California, Berkeley.

National Interagency Fire Center, Fire Investigation Team. "Cerro Grande Prescribed Fire, May 4–8, 2000, Investigation Report." 18 May 2000, http://www.nps.gov/cerrogrande/.

California

California Department of Natural Resources, Division of Forestry. *California Government and Forestry: From Spanish Days until the Creation of the Department of Natural Resources in 1927,* by C. Raymond Clar. Sacramento: California State Printing Office, 1959.

California Department of Public Works (CDPW). "Report of Consulting Board on Safety of the Proposed San Gabriel Dam, Los Angeles County." November 1929. Water Resources Center Archives, University of California, Berkeley.

———. *South Coastal Basin.* Bulletin No. 32. San Francisco: California State Printing Office, 1930.

California Department of Public Works (CDPW), Division of Highways. "Joint Departmental Report on Performance of Revetment and Bank Protection Subsequent to 1937–38 Floods in Northern and Southern California." 1938. Government Publications Section, California State Library, Sacramento, California.

California Department of Public Works (CDPW), Division of Water Rights. *Biennial Reports.* 1925–1927. Sacramento: California State Printing Office, 1925–1927.

———. *San Gabriel Investigation: Analysis and Conclusions,* by Harold Conkling. Bulletin No. 7, Reports of the Division of Water Rights. Sacramento: California State Printing Office, 1929.

———. *San Gabriel Investigation: Report for the Period July 1, 1923 to September 30, 1926.* Bulletin No. 5, Reports of the Division of Water Rights. Sacramento: California State Printing Office, 1927.

———. *San Gabriel Investigation: Report for the Period October 1, 1926 to September 30, 1928,* by Harold Conkling. Bulletin No. 6, Reports of the Division of Water Rights. Sacramento: California State Printing Office, 1929.

California State Board of Agriculture. *Statistical Report of the California State Board of Agriculture for the Year 1914.* Sacramento: California State Printing Office, 1915.

California State Legislature, Senate, Committee on Local Government. "Flood Control in the Antelope Valley: Organization and Financing: Summary of the Testimony Received at the Interim Hearing of the Senate Committee on Local Government." 30 July 1986, Lancaster, California.

California State Planning Board. "Flood-Plain Zoning: Possibilities and Legality with Special Reference to Los Angeles County, California," by Ralph B. Wertheimer. June 1942. Water Resources Center Archives, University of California, Berkeley.

City and County

City of Los Angeles, Los Angeles River Task Force. "Report of the City of Los Angeles, Los Angeles River Task Force." January 1992. Los Angeles Public Library, Los Angeles.

Denver Urban Renewal Authority. "Economic Feasibility Study for the Denver Platte River Valley," by Reuben A. Zubrow. 7 June 1966. University of Colorado, Boulder.

Los Angeles County Board of Engineers Flood Control (Board of Engineers). "Provisional Report of Board of Engineers Flood Control Submitted to the Board of Supervisors Los Angeles County." 1914. Rare Books Collection, Huntington Library, San Marino, California.

———. "Report on San Gabriel River Control," by Frank H. Olmsted. 6 October 1913. Rare Books Collection, Huntington Library, San Marino, California.

———. *Reports of the Board of Engineers Flood Control to the Board of Supervisors Los Angeles County California.* Los Angeles, 1915. Rare Books Collection, Huntington Library, San Marino, California.

Los Angeles County Department of Public Works (LACDPW). "Hydrologic Report." 1994–96. Los Angeles County Department of Public Works Technical Library, Alhambra, California.

———. *Los Angeles River Alternative Flood Control Study,* vol. 1: *Baseline Conditions Report,* prepared by Simons, Li & Associates, Inc. Los Angeles, August 1997. Los Angeles Public Library, Los Angeles.

———. *Los Angeles River Alternative Flood Control Study,* vol. 2: *Evaluation of Alternatives,* prepared by Simons, Li & Associates, Inc. Los Angeles, August 1997. Los Angeles Public Library, Los Angeles.

Los Angeles County Department of Public Works, Department of Parks and Recreation, and Department of Regional Planning (LACDPW). "Los Angeles River Master Plan." June 1996. Los Angeles Public Library.

Los Angeles County Flood Control District (LACFCD). *Annual Reports.* 1930–

31, 1941–42 to 1958–59. Los Angeles, 1930–1959. Water Resources Center Archives, University of California, Berkeley.
———. "Antelope Valley: Report by the Los Angeles County Engineer and the Los Angeles County Flood Control District Concerning Flood Control and Water Conservation," by A. E. Bruington and J. A. Lanbie. October 1970.
———. "Benefit Assessment for Flood Control." July 1979. Water Resources Center Archives, University of California, Berkeley.
———. *Biennial Reports.* 1959–61 to 1981–83. Los Angeles, 1959–1983. Water Resources Center Archives, University of California, Berkeley.
———. "Comprehensive Plan for Flood Control and Conservation: Immediate Needs." Map approved by E. Courtlandt Eaton. N.d. Los Angeles Department of Public Works Technical Library, Alhambra, California.
———. "Comprehensive Plan for Flood Control and Conservation: Present Conditions and Immediate Needs." September 1931. Water Resources Center Archives, University of California, Berkeley.
———. "Comprehensive Plan for Flood Control and Water Conservation Assembled from Computations and Data Made prior to February 18, 1935." 1935. Los Angeles County Department of Public Works Technical Library, Alhambra, California.
———. "Comprehensive Plan: Preliminary Layout of Ultimate Development." Map approved by E. Courtlandt Eaton. N.d. Los Angeles County Department of Public Works Technical Library, Alhambra, California.
———. "Comprehensive Plan: Preliminary Outline of Ultimate Plan." 10 September 1930. Los Angeles County Department of Public Works Technical Library, Alhambra, California.
———. "Conditions at District Dams Following Flood of March 2, 1938," by Finley B. Laverty and N. B. Hodgkinson. 21 March 1938. Los Angeles County Department of Public Works Technical Library, Alhambra, California.
———. "Evaluation of Alternatives for Debris and Flood Control in Rustic and Rivas Canyons." October 1984. Los Angeles Department of Public Works Technical Library, Alhambra, California.
———. "Flood of March 2, 1938," by M. F. Burke. 20 May 1938. Water Resources Center Archives, University of California, Berkeley.
———. "Floods of January 15–18, 1952," by M. F. Burke. August 1952. Water Resources Center Archives, University of California, Berkeley.
———. *Homeowner's Guide for Debris and Erosion Control.* N.d. Water Resources Center Archives, University of California, Berkeley.
———. *Hydrology Manual.* Los Angeles: Los Angeles County Flood Control District, 1971.
———. "Los Angeles River Flood Prevention Project Review Report," by Earl C. Ruby. 1974. Los Angeles Department of Public Works Technical Library, Alhambra, California.
———. *1983 Storm Report.* Los Angeles, 1983. Water Resources Center Archives, University of California, Berkeley.
———. "Preliminary Report on Foothill Flood of December 31st 1933." 4 Jan-

uary 1934. Los Angeles Department of Public Works Technical Library, Alhambra, California.

———. "Proposed Modification of the Design of San Gabriel Dam No. 1." 1935. Rare Books Collection, Huntington Library, San Marino, California.

———. "Report of the Chief Engineer of the Los Angeles County Flood Control District Comparing the Dam and the Reservoir Site at San Gabriel Forks and the Dam and Reservoir Site at Pine Canyon Containing Comparative Maps Made from Actual Surveys and Other Data." 31 July 1925. Harriet Williams Russell Strong Collection, Huntington Library, San Marino, California.

———. *Report of J. W. Reagan Chief Engineer of Los Angeles County Flood Control District upon the Control and Conservation of Flood Waters in This District by Correction of Rivers, Diversion and Care of Washes, Building of Dikes and Dams, Protecting Public Highways, Private Property and Los Angeles and Long Beach Harbors.* Los Angeles, 1926. Los Angeles County Department of Public Works Technical Library, Alhambra, California.

———. *Report of J. W. Reagan Chief Engineer Los Angeles County Flood Control District upon the Control of Flood Waters in This District by Correction of Rivers, Diversion and Care of Washes, Building of Dikes and Dams, Protecting Public Highways, Private Property and Los Angeles and Long Beach Harbors.* Los Angeles, 1917. Rare Books Collection, Huntington Library, San Marino, California.

———. *Report of J. W. Reagan Chief Engineer of Los Angeles County Flood Control District, upon the Control of Flood Waters in This District by Correction of Rivers, Diversion and Care of Washes, Building of Dikes and Dams, Protecting Public Highways, Private Property and Los Angeles and Long Beach Harbors.* Los Angeles, 1924. Water Resources Center Archives, University of California, Berkeley.

———. *Report of S. M. Fisher Chief Engineer of the Los Angeles County Flood Control District on Control and Conservation of Flood, Storm or Other Waste Waters of the District.* Los Angeles, 1934. Water Resources Center Archives, University of California, Berkeley.

———. "Report on Debris Reduction Studies for Mountain Watersheds," by William R. Ferrell, W. R. Barr, K. D. Matthews, R. Nagel, and J. S. Angus. November 1959. Water Resources Center Archives, University of California, Berkeley.

———. "Report on Flood of January 21–23, 1943," by Finley B. Laverty. August 1943. Water Resources Center Archives, University of California, Berkeley.

———. "Review of San Gabriel Case before the Federal Commission," by James W. Reagan. 29 April 1926. Los Angeles County Department of Public Works Technical Library, Alhambra, California.

———. "Summary Report: Storms of 1969," by Larry D. Simpson. June 1969. Los Angeles Department of Public Works Technical Library, Alhambra, California.

———. "Tentative Report to the Board of Supervisors of the Los Angeles

County Flood Control District Outlining the Work Already Done and Future Needs of Flood Control and Conservation with Tentative Estimates, Maps, Plans and Flood Pictures," by James W. Reagan. 7 February 1924. James Barlow Lippincott Papers, Water Resources Center Archives, University of California, Berkeley, California.

———. "Water Supplies for Los Angeles County," by Stephen J. Koonce and A. E. Bruington. February 1977. Los Angeles Department of Public Works Technical Library, Alhambra, California.

LAWS AND COURT CASES

"An Act to Create a Flood Control District." *Statutes and Amendments to the Codes of California* (1915), chap. 755, 1502–12.

DeBaker v. Southern California Railway Company. 39 *Pac. Rep.* 615 (1895).

Los Angeles County Flood Control District v. Henry W. Wright. 213 *Cal. Rep.* 335 (1931).

INTERVIEWS

MacAdams, Lewis. Telephone interview by author. 15 May 2002.

Mui, Chi. Telephone interview by author. 15 May 2002.

MANUSCRIPT AND ARCHIVAL COLLECTIONS

Barnes, Carleton P. Papers. Water Resources Center Archives, University of California, Berkeley.

Barnes, Harry. Papers. Water Resources Center Archives, University of California, Berkeley.

California History Room (CHR). California State Library, Sacramento.

California Scrapbook. Huntington Library, San Marino, California.

Cooper, Erwin. Papers. Water Resources Center Archives, University of California, Berkeley.

Couts, Cave Johnson. Papers. Huntington Library, San Marino, California.

Elliott, Thomas Balch (TBE). Papers. Huntington Library, San Marino, California.

Ford, John Anson (JAF). Papers. Huntington Library, San Marino, California.

Galloway, John Debo. Papers. Water Resources Center Archives, University of California, Berkeley.

Government Publications Section (GPS). California State Library, Sacramento.

Hyatt, Edward. Papers. Water Resources Center Archives, University of California, Berkeley.

Lippincott, James Barlow (JBL). Papers. Water Resources Center Archives, University of California, Berkeley.

Los Angeles Area Chambers of Commerce Collection (LAACC). Regional History Center, University of Southern California, Los Angeles.

Los Angeles City Archives. Los Angeles.

Los Angeles County Department of Public Works Technical Library (LACDPWTL). Alhambra, California. (Miscellaneous historical documents.)

Los Angeles Public Library (LAPL), Los Angeles.
Lukens, Theodore Parker (TPL). Papers. Huntington Library, San Marino, California.
Lynch, Henry Baker (HBL). Papers. Department of Special Collections, Charles E. Young Research Library, University of California, Los Angeles (DSCUCLA).
Minute Books. Los Angeles County Board of Supervisors (LACBS), Executive Office, Los Angeles.
Nixon, Richard (RNL). Papers. Richard Nixon Library, Yorba Linda, California.
Nordskog, Andrae B. Papers. Water Resources Center Archives, University of California, Berkeley.
Old Document Files. Los Angeles County Board of Supervisors (LACBS), Executive Office, Los Angeles.
Oral History Program. Department of Special Collections, Charles E. Young Research Library, University of California, Los Angeles (DSCUCLA).
Rare Books Collection. Huntington Library, San Marino, California.
Rogers, J. David (JDR). Private Historical Collections. Pleasant Hill, California.
Sierra Club. Oral History Project. Bancroft Library, University of California, Berkeley, and Department of Special Collections, Charles E. Young Research Library, University of California, Los Angeles (DSCUCLA).
Sierra Club. Papers. Department of Special Collections, Charles E. Young Research Library, University of California, Los Angeles (DSCUCLA).
Sierra Madre City Council, Minutes of Meetings (SMCCM). City Hall, Sierra Madre, California.
Strong, Harriet Williams Russell (HWRS). Papers. Huntington Library, San Marino, California.
Thrall, William H. Papers. Huntington Library, San Marino, California.
U.S. Army Corps of Engineers. Papers. Water Resources Center Archives, University of California, Berkeley.
U.S. Army Corps of Engineers. Public Affairs Office (PAO), Los Angeles, California. (Miscellaneous historical documents.)
U.S. Army Corps of Engineers. Record Group 77. National Archives, Pacific Region (NAPR), Laguna Niguel, California.
U.S. Bureau of Land Management, General Land Office. Record Group 49. National Archives, Pacific Region (NAPR), Laguna Niguel, California.
Water Resources Center Archives (WRCA), University of California, Berkeley. (Miscellaneous historical documents.)
Willard, Charles Dwight (CDW). Papers. Huntington Library, San Marino, California.

PERIODICALS

Current News: The Voice of the River (Los Angeles), 1996–1999.
Engineering News-Record, 1924–1950.
Los Angeles Examiner, 1914.
Los Angeles Times, 1914–2002.
Municipal League of Los Angeles Bulletin, 1925–1941.

Index

Text:	10/13 Sabon
Display:	Sabon
Compositor:	G&S Typesetters, Inc.
Printer and Binder:	Thomson-Shore, Inc.